Selected Papers from IEEE ICASI 2019

Selected Papers from IEEE ICASI 2019

Special Issue Editors

Sheng-Joue Young
Shoou-Jinn Chang
Stephen D. Prior
Liang-Wen Ji

MDPI • Basel • Beijing • Wuhan • Barcelona • Belgrade

Special Issue Editors
Sheng-Joue Young
National Formosa University
Taiwan

Shoou-Jinn Chang
National Cheng Kung University
Taiwan

Stephen D. Prior
University of Southampton
UK

Liang-Wen Ji
National Formosa University
Taiwan

Editorial Office
MDPI
St. Alban-Anlage 66
4052 Basel, Switzerland

This is a reprint of articles from the Special Issue published online in the open access journal *Applied Sciences* (ISSN 2076-3417) from 2019 to 2020 (available at: https://www.mdpi.com/journal/applsci/special_issues/icasi2019applsci).

For citation purposes, cite each article independently as indicated on the article page online and as indicated below:

LastName, A.A.; LastName, B.B.; LastName, C.C. Article Title. *Journal Name* **Year**, *Article Number*, Page Range.

ISBN 978-3-03928-929-5 (Hbk)
ISBN 978-3-03928-930-1 (PDF)

Contents

About the Special Issue Editors

Sheng-Joue Young received his bachelor's degree from the Department of Physics, National Changhua University of Education, Changhua, Taiwan, in 2003. He received his M.S. and Ph.D. degrees from the Institute of Electro-Optical Science and Engineering and Institute of Microelectronics, National Cheng Kung University (NCKU), Taiwan, in 2005 and 2008, respectively. He was a visiting research student with the Graduate School of Engineering, Chiba University, Chiba, Japan, from July to September 2007. In 2010, he became an Associate Professor with the Department of Electronic Engineering, National Formosa University, Yunlin, Taiwan. He was promoted to Associate Professor in August 2013. He is also a distinguished Professor with the Institute of Creative Industries Research, Fuzhou University, China, in 2014. He guided the institute to conduct academic research. His current research interests include semiconductor physics, optoelectronic devices, and nanotechnology. He has published more than 110 SCI journal papers.

Shoou-Jinn Chang received a B.S. degree from the National Cheng Kung University (NCKU), Tainan, Taiwan, in 1983; a M.S. degree from the State University of New York, Stony Brook, NY, USA, in 1985; and a Ph.D. from the University of California, Los Angeles, CA, USA, in 1989, all in electrical engineering. From 1989 to 1992, he was a Research Scientist with Nippon Telegraph and Telephone Basic Research Laboratories, Musashino, Tokyo, Japan. He joined the Department of Electrical Engineering, NCKU, in 1992 as an Associate Professor, where he was promoted to Full Professor in 1998. He served as the Director of the Institute of Microelectronics, NCKU, from August 2008 to July 2011, and the Deputy Director of the Center forMicro/Nano Science and Technology, NCKU, from February 2006 to January 2011. His current research interests include semiconductor physics, optoelectronic devices, and nanotechnology. Dr. Chang received the Outstanding Research Award from the National Science Council, Taiwan, in 2004. He is a Fellow of the Optical Society of America (OSA), the International Society for Optical Engineers (SPIE), and the Institute of Electrical and Electronics Engineers (IEEE).

Stephen D. Prior is a drone expert, engineer, and educationalist who leads a multidisciplinary research group at the University of Southampton, USA. His innovative work on drone design builds on his personal experience as an engineer and scientist. Stephen has built an unorthodox and creative team of aeronautical engineers, design engineers, and robotic experts. Stephen has an international profile as an academic and conference chair. He publishes widely and speaks frequently at national and international events. He is especially interested in collaborative research at the intersections between traditional disciplinary boundaries and brings his considerable knowledge and insight to thinking on unmanned systems and the future of aircraft.

Liang-Wen Ji received a B. S. degree in physics, a M.S. degree in material science, and a Ph.D. degree in electrical engineering from the National Cheng Kung University (NCKU), Tainan, Taiwan. In 2005, he became an Associate Professor with the Institute of Electro-Optical and Materials Science, National Formosa University (NFU), Yunlin, Taiwan, and was promoted to Full Professor in 2007. From August 2005 to July 2006, he served as the Director of the R&D Center for Flat Panel Display Technology, NFU. His current research interests include semiconductor physics, optoelectronic devices, and nanotechnology. Dr. Ji was the recipient of the Research Award from Lam Research Taiwan Co., Ltd., Taiwan, in 2004.

applied sciences

MDPI

Editorial

Special Issue: Selected Papers from IEEE ICASI 2019

Sheng-Joue Young [1,*], Shoou-Jinn Chang [2], Stephen D. Prior [3] and Liang-Wen Ji [4]

1. Department of Electronic Engineering, National Formosa University, Yunlin 632, Taiwan
2. Department of Electrical Engineering, National Cheng Kung University, Tainan 701, Taiwan; changsj@mail.ncku.edu.tw
3. Aeronautics, Astronautics and Computational Engineering, University of Southampton, Southampton SO16 7QF, UK; S.D.Prior@soton.ac.uk
4. Department of Electro-Optical Engineering, National Formosa University, Yunlin 632, Taiwan; lwji@nfu.edu.tw

* Correspondence: shengjoueyoung@gmail.com or youngsj@nfu.edu.tw

Received: 25 March 2020; Accepted: 27 March 2020; Published: 11 April 2020

1. Introduction

The fifth IEEE International Conference on Applied System Innovation 2019 (IEEE ICASI 2019, https://2019.icasi-conf.net/), which was held in Fukuoka, Japan, on April 11–15, 2019, provided a unified communication platform for a wide range of topics. This Special Issue titled "Selected Papers from IEEE ICASI 2019" collected nine excellent papers presented on the applied sciences topic during the conference. Mechanical engineering and design innovations are academic and practical engineering fields that involve systematic technological materialization through scientific principles and engineering designs. Technological innovation by mechanical engineering includes information technology (IT)-based intelligent mechanical systems, mechanics and design innovations, and applied materials in nanoscience and nanotechnology. These new technologies, which implant intelligence in machine systems, represent an interdisciplinary area that combines conventional mechanical technology and new IT. The main goal of this Special Issue is to discover new scientific knowledge relevant to IT-based intelligent mechanical systems, mechanics and design innovations, and applied materials in nanoscience and nanotechnology.

2. Topics of Selected Papers

This Special Issue selected nine excellent papers from a total of 200 papers presented at IEEE ICASI 2019. The published papers are introduced as follows.

Dr. Lin [1] reported on the "Application of a Fuzzy Decision Model to the Design of a Pillbox for Medical Treatment of Chronic Diseases". This study describes the importance of reducing waste from resources in medical systems. Patients with chronic diseases generate medical waste because they often forget to take their medications. Patients pay attention to the time and amount of medicine to take at different doses. This negligence often affects the schedule of taking medicines or preventive drugs. The amount of medicine also varies for different patients. The evaluation model in this study utilizes the fuzzy analytic hierarchy process to obtain the degree (weight) of each evaluation item determined by each patient. Thereafter, a more objective overall shape can be determined depending on the individual's preference. In this study, the proposed package design serves as the case study. The results indicate that the evaluation model is feasible, and the findings of the case study are also valuable for follow-up designers. The design of a pillbox for patients with chronic diseases should consider the safety of taking the medicine; that is, the right person should take the right medicine at the right time at the right dose. Therefore, evaluating the feasibility of the pillbox for patients with chronic diseases is extremely important. The proposed evaluation model applies to products that have different compositions. Follow-up researchers or designers can apply this approach to different case designs.

Dr. Ko [2] reported on the "Application of Fuzzy Theory to the Evaluation Model of Product Assembly Design and Usability Operation Complexity". A product usually consists of more than one component to improve convenience and enhance living standards. However, a product is created by the joint efforts of people from various territories; therefore, an important consideration is enabling a product to merge into consumers' daily lives rather than simply fulfilling its functions. Conflicts may arise in people's life patterns or values, which should be considered during the manufacturing process. This study investigates the assembly process by considering the assembly operations and operating time. By determining the relationship between components, we analyzed the assembly concept of most components. A fuzzy comprehensive evaluation was conducted during the evaluation of the degree of complexity of user operations. Based on the ranking of membership, the most appropriate assembly was determined, which serves as a reference for designers to select the optimal product assembly. By recording the consumer usage models, we also proposed the optimal assembly and usage model of product design. The goal of this study is to find a balance between assembly evaluation and the usage model by allowing designers to determine the new assembly concepts that meet consumer usage models. A case study of four bedside stereos was conducted by implementing the proposed approach to determine the evaluation principle of assembly. The purpose is to enhance the balance between assembly design and user operation complexity for making efficient decisions. Thus, a product design can comply with the concept of concurrent engineering and the quality of a product design can be enhanced.

R. H. Hwang et al. [3] reported on "An LSTM-based Deep Learning Approach for Classifying Malicious Traffic at the Packet Level". Recently, deep learning has been successfully applied to network security assessments and intrusion detection systems (IDSs) with various breakthroughs, such as using convolutional neural networks and long short-term memory (LSTM) to classify malicious traffic. However, these state-of-the-art systems also face tremendous challenges to satisfy real-time analysis requirements due to the major delay of flow-based data preprocessing, i.e., requiring time to accumulate packets into particular flows and then extracting features. If detecting malicious traffic can be performed at the packet level, then the detection time will be significantly reduced, thereby improving the potential of online real-time malicious traffic detection based on deep learning technologies. To accelerate the entire detection process by considering a packet-level classification, which has not been studied in the literature, we propose a novel approach in building the malicious classification system with the primary support of word embedding and an LSTM model. Specifically, we propose a novel word-embedding mechanism to extract packet semantic meanings and adopt LSTM to learn the temporal relation among fields in the packet header and further classify whether an incoming packet is normal or a part of malicious traffic. The evaluation results on ISCX2012, USTC-TFC2016, the IoT dataset from Robert Gordon University, and the IoT dataset collected on our Mirai botnet show that our approach is competitive to the prior literature, which detects malicious traffic at the flow level. As network traffic continues to grow every year, our first attempt can inspire the research community to exploit the advantages of deep learning to build effective IDSs without suffering significant detection delay.

Zhu Chengtao et al. [4] reported on "Hierarchical Guided Image Filtering for Efficient Stereo Matching". Stereo matching is complicated by the uneven distribution of textures on image pairs. We address this problem by applying edge-preserving guided image filtering (GIF) at different resolutions. In contrast to most multi-scale stereo-matching algorithms, the parameters of the proposed hierarchical GIF model are in an innovative weighted-combination scheme to generate an improved matching cost volume. Our method draws its strength from exploiting texture in various resolution levels and performing an effective mixture of the derived parameters. This novel approach advances our recently proposed algorithm, the pervasive GIF scheme, by equipping it with hierarchical filtering modules, thereby resulting in a disparity in images with a large number of details. The approach ensures as many different-scale patterns as possible to be involved in the cost aggregation, thereby improving matching accuracy. The experimental results show that the proposed scheme achieves the best matching accuracy when compared with six well-recognized cutting-edge algorithms using version 3 of the Middlebury stereo evaluation data sets.

W. C. Lee et al. [5] reported on "Automatic Error Compensation for Free-form Surfaces by Using On-machine Measurement Data". Currently, most computer numerical control controllers lack the function needed to compensate machining errors for free-form surfaces. The objective of this study is to enhance the accuracy and precision of the machined free-form surfaces of a workpiece using a mirror compensation method with on-measurement data. A new free-form surface for finishing machining can be automatically reconstructed by mirroring the points measured after semi-finishing. The surface can then be used to generate the cutting tool path to reduce errors during finishing. In this study, three types of surfaces were used to evaluate the proposed method. The results show that the proposed method reduced the standard deviations of the three surface geometries by 61%, 61%, and 32%. We also evaluated the tool radius modification method commonly used in the industry for error compensation and found no substantial reduction on standard deviation. Therefore, the effectiveness of the proposed error compensation method is evident.

Y. C. Hung et al. [6] reported on "3D and Boundary Effects on 2D Electrical Resistivity Tomography". Electrical resistivity tomography (ERT) is one of the most widely used geophysical methods in geological, hydrogeological, and geo-environmental investigations. Although 3D ERT is now available, 2D ERT remains state-of-the-practice due to its simplicity in fieldwork and lower space requirements. Two-dimensional ERT assumes that the ground condition is perpendicular to the survey line and homogeneous outside the survey line. This assumption can often be violated in conditions such as geologic strikes not being perpendicular to the survey line and topographic changes or buried objects near the survey line. Possible errors or artifacts in the 2D resistivity tomogram arising from violating the 2D assumption are often overlooked. This study aimed to numerically investigate the boundary effects on 2D ERT under various simplified conditions. Potential factors including resistivity contrast, the depth and size of buried objects, and electrode spacing were considered for the parametric studies. The results reveal that offline geologic features may project onto the 2D tomogram to a certain extent, depending on the aforementioned factors. The mechanism and implications of boundary effects can be drawn from these parametric studies.

S. N. Tang et al. [7] reported on "Area-efficient FFT Kernel with Improved Use of GI for Multistandard MIMO-OFDM Applications". This study presents a fast Fourier transform (FFT) kernel for multistandard applications, which employ multiple-input, multiple-output orthogonal frequency-division multiplexing (MIMO-OFDM). The proposed design uses a mixed-radix, mixed-multipath delay-feedback (MRM2DF) structure, which enables 4/5/6-stream 64/128-point FFT. This approach allows for the effective usage of guard intervals in conjunction with a novel resource-sharing scheme to improve area efficiency. An area-reduced constant multiplication unit and sorting buffer with minimal memory size further reduced an area overhead. A test chip was designed using UMC 90-nm technology and was evaluated through post-layout simulation. The proposed design outperformed previous works in terms of throughput per area.

M. C. Hwang et al. [8] reported on "The Design and Building of a Hexapod Robot with Biomimetic Legs". A hexapod robot with biomimetic legs was built to implement a distributed control system, where a mechanism is proposed to serve as the central pattern generator and a computer to act as the brain stem, cooperating with the central pattern generator through wireless communication. The proposed mechanism consists of two modules, i.e., tripod gait generator and Jansen's linkage. The tripod gait generator is a device that uses a single motor to generate a tripod gait, while Jansen's linkage rhythmically executes the legged motion. In a sense, we are trying to implement the locomotion of a robot by means of a hybrid computational system, including the mechanism and electronic processor parts. The complex mathematical function of the foot movement is realized by the ensemble of links of the Jansen's linkage to alleviate the computational burden. Furthermore, the proposed design, based on non-collocated actuators, is intended to minimize the number of actuators while reducing the building cost of the robot.

C. M. Lin et al. [9] reported on "Heating, Ventilation, and Air Conditioning System Optimization Control Strategy Involving Fan Coil Unit Temperature Control". The objective of this study was to develop a heating, ventilation, and air conditioning (HVAC) system optimization control strategy involving fan coil unit (FCU) temperature control for energy conservation in chilled water systems to enhance the operating efficiency of HVAC systems. The proposed control strategy involves three techniques, which are described as follows. The first technique is an algorithm for dynamic FCU temperature setting, which enables the FCU temperature to be set in accordance with changes in the outdoor temperature to satisfy the indoor thermal comfort of occupants. The second technique aims to determine the indoor cold air demand, which collects the set FCU temperature and converts it to the refrigeration ton required for the chilled water system, thereby serving as the control target for ensuring optimal HVAC operation. The third technique is a genetic algorithm for calculating the minimum energy consumption for an HVAC system. The genetic algorithm determines the pump operating frequency associated with minimum energy consumption per refrigeration ton to control energy conservation. A field experiment was conducted to demonstrate the effectiveness of the proposed HVAC system optimization control strategy combining FCU temperature control. The results show that the proposed strategy enabled the HVAC system to achieve 39.71% energy conservation compared with the system operating at full load.

Acknowledgments: The guest editors would like to thank the authors for their contributions to this Special Issue and all the reviewers for their constructive comments. We are also grateful to Xiaoyan Chen, Assistant Managing Editor of Applied Sciences, for the time and efforts she contributed toward the publication of this Special Issue.

Conflicts of Interest: The authors declare no conflict of interest.

References

1. Lin, H.-H. Application of a Fuzzy Decision Model to the Design of a Pillbox for Medical Treatment of Chronic Diseases. *Appl. Sci.* **2019**, *9*, 4909. [CrossRef]
2. Ko, Y.-C. Application of Fuzzy Theory to the Evaluation Model of Product Assembly Design and Usability Operation Complexity. *Appl. Sci.* **2019**, *9*, 4055. [CrossRef]
3. Hwang, R.H.; Peng, M.C.; Nguyen, V.L.; Chang, Y.L. An LSTM-Based Deep Learning Approach for Classifying Malicious Traffic at the Packet Level. *Appl. Sci.* **2019**, *9*, 3414. [CrossRef]
4. Zhu, C. Hierarchical Guided-Image-Filtering for Efficient Stereo Matching. *Appl. Sci.* **2019**, *9*, 3122. [CrossRef]
5. Lee, W.C.; Lee, Y.T.; Wei, C.C. Automatic Error Compensation for Free-Form Surfaces by Using On-Machine Measurement Data. *Appl. Sci.* **2019**, *9*, 3073. [CrossRef]
6. Hung, Y.C.; Lin, C.P.; Lee, C.T.; Weng, K.W. 3D and Boundary Effects on 2D Electrical Resistivity Tomography. *Appl. Sci.* **2019**, *9*, 2963. [CrossRef]
7. Tang, S.N.; Chen, Y.H. Area-Efficient FFT Kernel with Improved Use of GI for Multistandard MIMO-OFDM Applications. *Appl. Sci.* **2019**, *9*, 2877. [CrossRef]
8. Hwang, M.C.; Liu, F.; Yang, J.; Lin, Y. The Design and Building of a Hexapod Robot with Biomimetic Legs. *Appl. Sci.* **2019**, *9*, 2792. [CrossRef]
9. Lin, C.M.; Liu, H.Y.; Tseng, K.Y.; Lin, S.F. Heating, Ventilation, and Air Conditioning System Optimization Control Strategy Involving Fan Coil Unit Temperature Control. *Appl. Sci.* **2019**, *9*, 2391. [CrossRef]

applied sciences

MDPI

Article

Application of a Fuzzy Decision Model to the Design of a Pillbox for Medical Treatment of Chronic Diseases

Hsin-Hung Lin [1,2]

1 Department of Creative Product Design, Asia University, Taichung City 41354, Taiwan; hhlin@asia.edu.tw or a123lin0@gmail.com; Tel.: +886-04-2332-3456-1051
2 Department of Medical Research, China Medical University Hospital, China Medical University, Taichung 404, Taiwan

Received: 14 September 2019; Accepted: 7 November 2019; Published: 15 November 2019

Abstract: It is critical for medical systems to reduce waste from medical resources. One of the reasons why patients with chronic diseases create waste is that they often forget to take their medicine. Patients pay attention to the time and amount of medicine to take to different degrees. This negligence often affects when they take medicine or preventive drugs. The amount of medicine is also different for different patients. The evaluation model in this study utilizes the fuzzy analytic hierarchy process (FAHP) to obtain the degree (weight) of each evaluation item that is determined by each patient. After that, a more objective overall shape can be determined depending on the individual's preference. In this study, the proposed package design serves as the case study. The results indicate that the evaluation model is feasible, and the research results of the case study are also valuable for follow-up designers. The design of a pillbox for patients with chronic diseases should consider the safety of taking the medicine; that is, the right person should take the right medicine at the right time at the right dose. Therefore, evaluating the feasibility of the pillbox for patients with chronic diseases is very important. The proposed evaluation model applies to products that have different compositions. Follow-up researchers or designers can apply this approach to different case designs.

Keywords: product design; layout strategy; design method

1. Introduction

The idea of a conceptual design phase was proposed by Pahl and Beitz in Germany and it has been viewed as the cornerstone of any conceptual design task. Their book has been used as the textbook for engineering design courses and implementation. Their approach was extended by introducing new technologies and tools. These technologies and tools can be embodied, and the extended design approach is called the integrated, customer driven, conceptual design method (ICDM). This approach can be applied to the entire conceptual design phase from defining demands to determining the optimal concept. Three out of the seven new ICDM tools can be used to enhance the task definition of quality function deployment (QFD), the conceptual failure mode analysis method (CFMA), and the design quality method (DQM) for assessing the satisfaction of the concepts that are generated. In addition to the Pahl and Beitz approach, ICDM also includes several steps that can contribute to the excellent design if they are fully implemented [1]. Gielisch proposed and evaluated four different product development approaches, which include the systematic approach of Pahl and Beitz (SAPB), theory of inventive problem solving (TRIZ), concurrent engineering (CE), and lean development (LD). These four approaches are specifically for the demand of micro-assembly and production on reconfigurable manufacturing systems (RMS). They feature clear structure, high integrated production, manufacturability, and time consumption. A new approach was proposed to successfully integrate the advantages of these four existing methods. The formal composition of an abstract product contains three data types. By using these attributes and a decision tree, the process that is required by RMS can be accomplished [2]. Fiorineschi proposed an approach to resolve the known deficiencies that can be contributed to systematic conceptual design (SCD). More precisely speaking, the well-known functional decomposition and morphology can be used to support creativity and to resolve the problem during the creation of innovative ideas during the activities of the SCD process. This approach is the problem–solution network that was developed in recent years. It provides a comprehensive guidance for the application of the TRIZ tool in SCD. In addition to their claim of resolving the deficiencies of functional decomposition and morphology (FDM), their study also considered PSN(Problem-Solution Network) for resolving problems and evolving into TRIZ integration. Their studies also proposed the application procedures to industrial case studies and carried out analyses in order to demonstrate their purpose [3]. Becattini applied general theory of powerful thinking (OTSM) and the theory of inventive problem solving to form an OTSM–TRIZ approach for students to resolve design problems within a problem network. This approach can be used to analyze design theories and design protocols. The purpose is to support the evaluation of students' overall achievement and to create systematic design courses for download learning. This approach also considered the process of generating new ideas and it allows people to carry out protocol studies during a limited amount of time and therefore it is very time-consuming [4]. During the Lorenzo integration process, the benefits of these three indicators can be utilized simultaneously. Innovative data can also be extracted from these three indicators during the evaluation process [5]. Jenab proposed a useful decision-making tooling for resolving the complexity of solutions during the conceptual design (CD) phase during the product development process. The purpose was to assess the potential new product investments so that the importance of the development strategy can be verified. The limit of the compatible information between design concepts still presents a challenge to the concept-selecting approaches. The layers within the decision standards can be aggregated to a chart, which can be used to determine the optimal alternative solution when evaluating the CD alternative [6]. The Erden universal framework can be used to understand functional modeling (FM). This approach highlights the characteristics of the FM-related classical methods so that the fundamental thinking can be classified [7]. Eckert proposed an approach for industrial designers, and it can be viewed as a portion of the quality function deployment (QFD) or failure mode and effect analysis (FMEA), which are suitable for functional modeling. This approach has been applied to the entire territory and has passed the formal process of an organization. Although both studies proposed the use of integrated design approaches, they were actually applied to the production demand and to form the structure rather than designing comprehensive tasks [8]. Ulrich proposed that the tasks that were related to design and manufacturing follow a plan that regulates the development stages, from the potential customer requesting a quotation to the final delivery of products. A product's functions can determine many related activities that should be designed to install in a specified region. A product also operates under the definite conditions so that the assembly manufacturer can work together with end users to enhance product quality and performance [9]. From the standpoint of creativity, Toh studied the conceptual selection process of engineering students in order to determine the factors that are helpful to the selection of creative concepts. A design team of students can carry out quantitative and qualitative analyses on the data that are acquired from controlled experiments. The results of similar studies indicated that the design team of students highlighted the technical feasibility of a design during the selection and discussion period. This is also highlighted in engineering education [10].

According to statistics, there is only a very small portion of people who can take the right medicine at the right time at the right dose. The chance for a patient who is more than 65 years old to take wrong medicine is seven times higher than those who are less than 65 years old. For the aged population in Taiwan, most of the elderly people who are more than 65 years old have developed more than one chronic disease. As a result, the safety issue of medicine is very important. At the moment, the pillboxes that are available on the market lack an important function of examining whether a patient has taken the same medicine repeatedly or has taken medicine with interfering effects. In order to allow a patient to take medicine regularly without worries, a new pillbox should be created for patients with chronic diseases as a solution to this type of problem. Moreover, the safety issue of taking medicine is considered in order to enhance the feasibility of a pillbox design for patients with chronic diseases so that they can take the right medicine at the right time at the right dose. As people are more concerned about environmental protection and are pursuing the goal of reducing the impact of a product on the environment to a minimum level, a new approach that integrates multi-criteria decision making (MCDM). This new approach is a fast and objective tool that is based on derivation from evidence. Earlier studies indicated that the proposed method can be applied to the evaluation of environmental functions of alternative design cases [11]. A comprehensive product evaluation method was proposed for students of industrial design so that they can carry out an effective design evaluation even when they encountered fuzzy information during the design evaluation stage. In this study, the fuzzy analytic hierarchy process (AHP) was used to evaluate product designs and to resolve the potential problems that might occur during the product evaluation process. The questionnaire with purposive selection was based on the judgment from experts so that different weights can be assigned to each evaluation criterion. The performance indicators of a fan were verified by CFD simulation and wind tunnel measurements. An optimal product can be created for evaluation of its optimized air flow rate. The integration of these methods is not only practical and objective, but also can help students make a decision between complicated and uncertain conditions. This approach also provides students with a good reference during the follow-up stage of product design and evaluation so that the learning competitiveness of students can be enhanced [12]. Similar to the technique for order of preference by similarity to ideal solution (TOPSIS), another framework was also proposed by integrating the analytical hierarchy process (AHP) with order priority technology. This approach can assist designers in identifying customer demands and design features. It is also helpful for determining the final design solution and for carrying out effective assessments. The proposed solution was first used in combination with the AHP for assessing customer demands and the overall relative importance of design features [13]. Fuzzy mathematics was proposed by the American control theory expert L.A. Zadeh in 1965 [14–16]. After years of development, the theoretical level is very comprehensive and this approach has been widely used in various areas, such as natural science and social science. Since fuzzy mathematics is used to handle objective and practical problems by its integration with precise mathematics, it is different from precise mathematics' characteristics of "one or the other" or "one and the other." Therefore, this approach finds a reason for its extensive applications. Moreover, the mathematical characteristics of the fuzzy approach also make it an emerging science within the design field. When assessing the quality of a lighting fixture design, for example, various factors must be considered. However, assessing the quality of a factor often includes human subjectivity and uncertainty, which leads to fuzziness. If an assessed result lacks objectivity, then reliability and validity will not be enhanced. Fuzzy mathematics can handle the condition of true and false and is also a feasible approach for measuring ideas. Whenever human thinking is involved, the fuzzy point-of-view could be a feasible approach for comprehensive assessment. Fuzzy theory can be applied to the selection of sites and studies, also indicating that the calculation of fuzzy numbers can also involve the ranking of several schemes based on their quality so that the quality of strategic analysis can be enhanced. In other studies, the fuzzy evaluation method is used to assess (computer-assisted instruction) the CAI curriculum. Fuzzy set theory is used to construct a multilevel fuzzy combined evaluation method for architecture colors. The goal is to determine the conclusion on the quality of the strategy of architectural color schemes. Kuo et al. [17] studied the selection of system operations of mobile value-added services by fuzzy theory. The results indicated that this model can effectively reduce human subjective awareness and enhance the precision of the selection so as to serve as the basis for the development of shape ideas. When a product design comes out, the fuzzy synthetic evaluation method can be used to assess product quality. The results indicate that building up this model can assist in the rationalization of the design process, reduce the misplay in the assessment, enhance the efficiency of the strategic decision, and increase the success rate of new product development [17].

Sun et al. [18] proposed applying triangular fuzzy numbers that cannot increase the reliability and authenticity of linear assessment by human sensory capture. Questionnaires are conducted in a qualitative manner to understand consumer preferences and needs through self-subjective judgments, matching collected terms, and screening consumers' shapes and functions for self-framed image vocabulary and expectation psychology. Then, using the quantization method, triangulation is used to calculate the ideal preference for the vocabulary closest to the consumer in different self-species. In their research, Hu and Liao studied 18-to-38-year-old individuals at colleges and universities, and then discussed various lifestyle attitudes and emotional consumption patterns of consumer groups. Consumer groups and frame styles have different factors to consider facets by evaluating various attributes of the vocabulary. By grouping consumers, they can understand the consumer groups and the consumption concepts they belong to and clearly understand their own positioning. Through the guidance of emotional imagery, consumers can find a path that suits them [19,20].

Fuzzy theory is a science used to study and deal with fuzzy phenomena. It was originally proposed by the computer scientist L.A. Zadeh of the University of California in 1965. It has been around for more than 40 years. It is used to process inaccurate and fuzzy data and operates through rigorous mathematical methods to solve decision problems in a fuzzy environment [21–23]. There are many objectives in the design evaluation, such as aesthetics (overall effect, shape, color, decoration, etc.), pleasure, safety, processability, etc., which are difficult to achieve by traditional quantitative analysis methods. To this end, it is necessary to introduce linguistic variables to describe and solve problems, and then use fuzzy mathematics to quantify the fuzzy information for quantitative evaluation [16].

Hsiao used fuzzy theory and analytic hierarchy analysis to make product decisions [14,24], and used fuzzy theory to perform monochrome color matching in the automotive design phase for related evaluations. In addition, he used fuzzy semantics to make automotive design decisions and lexical transformation product design descriptions to evaluate computer-aided industrial designs and images.

In addition, some scholars have proposed computer-aided systems to help design engineers reduce development and file processing time. Moskowitz and Kim proposed an optimal product design decision-support system in 1997 [15]. Developing an annual reasoning structure can infer the relationship between customer demand (CR) and design requirements (DR) [25]. However, developing these systems requires expertise and experience to build the rules while facing issues such as whether the system is performing well. According to fuzzy set theory, Kim et al. used supplier competition analysis to construct a relationship function between CR and DR and proposed a fuzzy multi-objective model. However, constructing these relationship functions has its difficulties, especially when developing a brand-new product. Data from competitors can be used for analysis [18]. Other scholars apply fuzzy sets, fuzzy operations, or defuzzification techniques to deal with complex and inaccurate quality functional problems; however, these methods do not consider the relationships between engineering design requirements [12,26]. Other scholars have emphasized that in addition to determining the degree of DR implementation based on customer satisfaction, we should also consider organizational conditions such as cost factors and technical difficulties to make economic, customer-friendly best decisions.

2. Theoretical Background

2.1. Fuzzy Theory

Fuzzy theory is a subject of science that deals with the study and processing of fuzzy phenomena. It was proposed in 1965 by L.A. Zadeh, an expert in control theory at the University of California, Berkeley, USA. It has been a subject for more than 40 years and has been used for processing imprecise data with vagueness. With rigorous mathematical methods for calculation, the strategic problems under a fuzzy environment can be resolved by this approach [14]. During design assessment, there are many targets, such as aesthetics (overall effect, shape, color, decoration, etc.), amenity, safety, machinability, etc. These targets are difficult to determine by conventional quantitative analysis. Therefore, linguistic variables must be introduced for description in order to acquire a solution. After that, the fuzzy mathematical approach can be used to numerate the fuzzy information for quantitative evaluation.

Hsiao utilized fuzzy theory and hierarchical analysis to carry out strategic decisions [25] and utilized fuzzy theory to evaluate monochromatic color schemes during the car appearance design phase [27,28]. In addition, he also applied fuzzy semantics to the strategic decision of automobile shape designs [21,22]. Moreover, the semantic transformation from adjectives was applied to the design of product forms in order to carry out computer-aided industrial design and assess form images [29,30].

In addition, a number of scholars have also proposed different computer-aided systems to assist designers and engineers in reducing the time for research and development and document processing. In 1997, Moskowitz and Kim proposed a decision support system with optimal product designs [31]. Temponi et al. proposed a deduction framework in 1999 in order to deduce the relationship between customer requirements (CRs) and design requirements (DRs) [32]. However, the knowledge and experience of experts are required to construct the rules when developing this type of system. At the same time, designers also encounter problems whether a system is in good operation or not. According to fuzzy set theory, Kim et al. utilized data, such as the analysis of the competitiveness of companies, to construct a correlation function between CRs and DRs. They proposed a fuzzy multi-objective model. However, it is difficult to build this correlation function. This is especially true when developing a brand-new product without any data from competitors for analysis [33]. Other scholars have applied fuzzy sets, fuzzy algorithms, or defuzzification techniques to process complex and imprecise quality function deployment problems. However, these methods do not take into account the relationships between engineering design demands. Other scholars have emphasized that the quality function deployment needs to not only determine the performance based on customer satisfaction but also consider organizational criteria such as cost factors and technical difficulties so as to determine the optimal decision that is economical and satisfactory for customers.

Among various targets of a design, some required concepts are definite while others are vague. It is also required to consider several targets and comprehensively consider various relevant factors and carry out comprehensive assessments. This is the so-called fuzzy comprehensive assessment. The process of implementing fuzzy assessment includes constructing the affecting factor set, determining the factor weight set, determining the parameter evaluation set, creating the single-factor assessment matrix, and conducting the fuzzy evaluation. These procedures are described as follows:

2.1.1. Constructing the Affecting Factor Set

When conducting a fuzzy evaluation, first confirm the factors affecting the values of the evaluation parameters. If it is known that the affecting factors are $u_1, u_2, \ldots, u_m$, the factor set that is composed of these parameters is $U = \{u_1, u_2, \ldots, u_m\}$. This factor set is a common set.

2.1.2. Determining the Factor Weight Set

The degree of influence and importance of each parameter for the factors are different; that is, the weight of each parameter is different for each factor. The factor weight set is the set that is composed of the degree of influence of each parameter for each factor and can be expressed as $A = \{a_1, a_2, \ldots, a_n\}$. If a_i indicates the ith factor weight, the weights of all of the factors need to satisfy Equation (1). A weight set is a fuzzy subset of the factor set and can be expressed as Equation (2):

$$\sum_{i=1}^{n} a_i = 1,\ a_i \geq 0\ (i = 1,2,3,\ldots,a_n), \tag{1}$$

$$\begin{aligned} A &= \tfrac{a_1}{u_1} + \tfrac{a_2}{u_2} + \tfrac{a_3}{u_3} + \ldots + \tfrac{a_n}{u_n} \\ &= \{a_1, a_2, a_3, \ldots, a_n\} \end{aligned} \tag{2}$$

The weight of each factor can be determined by the weight coefficient method, analytic hierarchical process, paired comparison method, Likert scale, or any subjective determination according to the practical problem. It does not matter which approach is used, and human factors are involved, with the only difference being in credibility. Here, A is the fuzzy set of each factor weight. For the same factor, if different data are taken, the assessment result is different. The method of the determinant table is to carry out pairwise comparison of the importance of these evaluation targets. Scores are given after further calculation and are entered into the table. The equation for calculating a_i is as follows:

$$a_i = k_i / \sum_{i=1}^{n} k_i, \tag{3}$$

where k_i is the total score of each evaluation target and n is the number of evaluation targets.

$$\sum_{i=1}^{n} k_i = \frac{n^2 - n}{2} \times 4 = 2(n^2 - n). \tag{4}$$

2.1.3. Determining the Parameter Evaluation Set

The evaluation set is composed of the various evaluation results that an assessor might obtain for the evaluation target. It is represented as V, and $V = \{v_1, v_2, \ldots, v_n\}$ and v_i $(i = 1, 2, 3, \ldots, n)$ represent the possible total evaluation results. The purpose of fuzzy evaluation is to comprehensively consider all of the affecting factors and obtain an optimal evaluation result from the evaluation set. The relationship between v_i and V is also a common set relationship. Therefore, the evaluation set is also a common set. For the evaluation in this study, the evaluation set is V = {*very satisfied, satisfied, neither satisfied nor dissatisfied, dissatisfied, very dissatisfied*}.

2.1.4. Creating the Single-Factor Assessment Matrix

A single-factor fuzzy evaluation is conducted to judge one factor separately and confirm the degree of membership for the target of evaluation toward evaluation-set elements. The evaluation target is carried out by the *i*th factor U_i and the membership grade of the *j*th element V_i in the evaluation set is r_{ij}. Therefore, the evaluation result of the *i*th factor U_i can be determined as follows:

$$R_i = \frac{r_{i1}}{V_1} + \frac{r_{i2}}{V_2} + \ldots + \frac{r_{in}}{V_n} \tag{5}$$

where Ri is called the single factor evaluation set, which is the fuzzy subset of the evaluation set. It can be expressed as $R_i = (r_{i1}, r_{i2}, \ldots, r_{in})$. Similarly, the single factor evaluation set of each factor can be determined as follows:

$$\begin{aligned} R_1 &= (r_{11}, r_{12}, \ldots, r_{1n}) \\ R_2 &= (r_{21}, r_{22}, \ldots, r_{2n}) \\ &\vdots \\ R_m &= (r_{m1}, r_{m2}, \ldots, r_{mn}) \end{aligned} \tag{6}$$

The fuzzy matrix that is composed of the membership grade of each single factor evaluation set is called the single-factor assessment matrix, R, as shown in Equation (7). R is a fuzzy matrix and can also be viewed as the fuzzy relational matrix from U to V, or so-called fuzzy mapping.

$$R = \begin{bmatrix} R_1 \\ R_2 \\ \vdots \\ R_i \\ \vdots \\ R_n \end{bmatrix} = \begin{bmatrix} r_{11} & r_{12} & \cdots & r_{1j} & \cdots & r_{1m} \\ r_{21} & r_{22} & \cdots & r_{2j} & \cdots & r_{2m} \\ \vdots & \vdots & \ddots & \vdots & \ddots & \vdots \\ r_{i1} & r_{i2} & \cdots & r_{ij} & \cdots & r_{im} \\ \vdots & \vdots & \ddots & \vdots & \ddots & \vdots \\ r_{n1} & r_{n2} & \cdots & r_{nj} & \cdots & r_{nm} \end{bmatrix}. \quad (7)$$

However, in this study, there are numerous factors to be considered and each factor can have different levels. It is difficult to resolve a problem by using single-factor fuzzy evaluation. It is also difficult to obtain reasonable assessment results. As a result, this study is based on fuzzy comprehensive evaluation, and then the fuzzy comprehensive evaluation is implemented. This is because there are too many factors to consider when making a selection from complicated schemes. There are levels between factors, therefore we must adopt the multifactor assessment matrix. The procedure is to divide the factor set into several levels according to its characteristics, carry out comprehensive assessment on each level, and then conduct in-depth combined evaluations on the evaluation results.

2.1.5. Conducting the Fuzzy Evaluation

If the fuzzy evaluation matrix of a certain scheme on the evaluation target is the R in Equation (7), the weighted comprehensive fuzzy evaluation and the product of fuzzy matrices to be considered is as follows:

$$B = A \bullet R = \left[b_1, b_2, \ldots b_j, \ldots b_m\right] \quad (8)$$

where • indicates the fuzzy synthetic operation.

There are several synthetic approaches of fuzzy weight matrix A and factor judgment matrix R. In this study, four synthetic approaches were implemented for the analysis and comparison of the evaluation results. These four models are described as follows:

Model 1: M (∧, ∨) synthetic operation, in which

$$b_j = \overset{m}{\underset{i=1}{\vee}} (a_i \wedge r_{ij}); j = 1, 2, \ldots, n, \quad (9)$$

where ∨ and ∧ indicate taking the maximum or minimum. When taking the minimum, the r_{ij} of all $r_{ij} > w_{ij}$ is not taken into consideration. Therefore, when there are various factors, the values are bound to be large after the weighted coefficients are normalized. As a result, a lot of the single-factor evaluation information will be lost. When there are fewer factors, w_{ij} might be larger and therefore the w_{ij} of all $w_{ij} > r_{ij}$ will not be taken into consideration. The influence of the main factors might be lost.

Model 2: M (•, ∨) synthetic operation, in which

$$b_j = \overset{m}{\underset{i=1}{\vee}} (a_i r_{ij}); j = 1, 2, \ldots, n. \quad (10)$$

The feature of this model deals with no loss of any useful information. However, the operation of taking the maximum ∨ could possibly lose a lot of useful information. It can still reflect the single-factor evaluation result and the degree of importance of each factor.

Model 3: M (∧, +°) synthetic operation, in which

$$b_j = \min\{1, \sum_{i=1}^{m} (a_i \wedge r_{ij})\}; j = 1, 2, \ldots, n, \quad (11)$$

where +° takes the summation with an upper limit of 1. The feature of this model deals with the loss of a lot of valuable information during the operation of taking the minimum. Therefore, it might not achieve the expected evaluation result. When the values of w_i and r_{ij} are larger, the resulting b_j could be equal to the upper limit of 1. When the values of w_i and r_{ij} are smaller, the resulting b_j could be equal to the summation of all w_i As a result, it could be more difficult to obtain the expected evaluation result.

Model 4: M $(\bullet, +)$ synthetic operation, in which

$$b_j = \min\{1, \sum_{i=1}^{m} a_i r_{ij}\}; j = 1, 2, \ldots, n. \tag{12}$$

This model is also called the weighted-average model and it deals with the situation when w_i is equipped with normalization, i.e., $\sum\limits_{i=1}^{m} a_i = 1$, $\sum\limits_{i=1}^{m} a_i r_{ij} << 1$.This model can be restructured as M $(\bullet, +)$, in which

$$b_j = \sum_{i=1}^{m} a_i r_{ij}; j = 1, 2, \ldots, n; \sum_{i=1}^{m} a_i = 1, \tag{13}$$

where $\sum\limits_{i=1}^{m} a_i = 1$. This model not only considers the influence of all factors, but also keeps all the information of the single-factor evaluation. During its operation, it does not apply the upper limit to a_i and r_{ij} $(i = 1, 2, \cdots, m; j = 1, 2, \cdots, n)$. However, a_i should be normalized. These are the significant characteristics and advantages of this model. When carrying out the fuzzy comprehensive evaluation and fuzzy optimization design on engineering design parameters, this model is implemented, since typically it can obtain better effects. This model not only considers the influence of all factors, but also keeps all information of the single-factor evaluation. During the operation, there is no upper limit on w_i and r_{ij}. It is only required to carry out normalization on w_i. This is the distinguishing feature and the main advantage of this model.

The goal of Models 1–3 is to obtain individual evaluation results based on certain limitations and by taking the extreme value. Therefore, a lot of useful information could be lost to various degrees during the evaluation process. As a result, these three models are applied to the scenarios that only care about the extreme values of objects in order to highlight certain main factors. Based on this, Model 4 is used for the synthetic operation in this study.

2.1.6. Processing the Evaluation Indices

After the evaluation indices b_j (j = 1, 2, ... , n) are determined, the results of the evaluation target can be determined by the methods of maximum degree of membership and weighted average, as follows.

(a) Maximum degree of membership

Based on the principle of maximum degree of membership, the evaluation element v_i corresponding to the maximum evaluation index b_j is selected. This approach considers only the contribution of the maximum evaluation index; information supplied by other indices is neglected. Moreover, when there are more than one maximum evaluation indices, it will be difficult for maximum degree of membership to determine a concrete result. In this case, weighted average is usually used.

(b) Weighted average

By taking b_j as the weight, the weighted average of each evaluation element v_j is determined to obtain the evaluation result; that is,

$$D = \frac{\sum\limits_{j=1}^{n} b_j v_j}{\sum\limits_{j=1}^{n} b_j}. \tag{14}$$

When the evaluation index b_j is normalized, then

$$D = \sum_{j=1}^{n} b_j v_j. \tag{15}$$

If the evaluation target is quantitative, Equation (14) can be used to determine the D value, which is the result of the fuzzy comprehensive evaluation for the evaluation target. If the target is not quantitative and the evaluation set is {superior, good, neither good nor bad, bad}, quantification of the nonquantitative targets of superior, good, neither good nor bad, and bad should be carried out. Otherwise, maximum degree of membership should be used. Based on these evaluation criteria, the distribution of various characteristics of the evaluation target can be determined. This method provides a deeper understanding of the evaluation target in order to determine the most appropriate processing approach to evaluate the data.

2.2. Analytic Hierarchy Process (AHP)

The analytic hierarchy process is a decision method proposed by Saaty (1980). It mainly applies to uncertain situations and multicriteria decision-making problems [34,35]. The AHP systemizes complicated multicriteria problems with a concise hierarchical framework. A decision maker carries out pairwise comparison between two criteria on their relative importance within a level. After that, a pairwise comparison matrix is constructed in order to determine the relative importance between the criteria. The overall prioritized vector of the entire level can be calculated by making the levels serially connected, leading to the weight of each evaluation criterion. The quantized result assists decision-makers in making comprehensive evaluations of alternative schemes in order to determine their priority and reduce the risk of making a wrong decision. Liu et al. proposed a new approach of calculating weights to replace the pairwise comparison of the analytic hierarchy process (AHP), called the voting analytic hierarchy process. This approach is simpler than AHP and can calculate weights systematically. It can also determine the priority of various suppliers by determining each one's score.

The procedure of the analytic hierarchy process includes five steps:

Step 1: Delimit the decision-making problem.

Step 2: Build the hierarchical framework.

Step 3: Construct the pairwise comparison matrix, as shown in Table 1, which is the assessment scale and relative comparison of the hierarchy analysis.

Table 1. Assessment scale and relative definitions of hierarchy analysis.

Scale	Definition
1	Equally important
3	Slightly important
5	Important
7	Very important
9	Absolutely important
2, 4, 6, 8	Intermediate values

Step 4: Calculate the eigenvalues.

Step 5: Examine the consistency.

When calculating the eigenvectors in Step 4, the following algorithms can be used:

(1) Theoretical analysis of eigenvalues and eigenvectors.

(2) Four approximate solutions that were proposed by Saaty: Normalization of the means of row vectors, also called normalization of the row average; normalization of the means of column vectors, also called the means of normalized columns; normalization of the geometric means of

row vectors, also called normalization of the geometric mean of the rows; and the average of the inverse of the normalized rows.

The consistency examination in Step 5 determines whether the evaluation results obtained from the pairwise comparison are consistent. In other words, it determines whether the experts' preference satisfies the transitivity. Saaty proposed the use of the consistency index (Consistency Index, *C.I.*) and consistency ratio (Consistency Ratio, *C.R.*) to carry out the examination. If both *C.I.* and *C.R.* are less than 0.1, it indicates that the pairwise matrix has consistency. The equations of *C.I.* and *C.R.* are as follows:

$$\text{C.I.} = \frac{\lambda_{\max} - n}{n-1} \tag{16}$$

where $\lambda_{\max}$ is the maximum eigenvalue of the matrix, and n is the order of the matrix (number of parameters); and

$$C.R. = \frac{C.I.}{R.I.} \quad \begin{aligned} C.R. &< 0.1 \Rightarrow \text{OK} \\ C.R. &= Consistencyratio \\ C.I. &= Consistencyindex \\ R.I. &= \text{Random } index \end{aligned} \tag{17}$$

where n is the number of evaluation criteria, and *R.I.* is the random index, and its value increases with the number of criteria, as shown in Table 2.

Table 2. Stochastic indicator table.

Order N	1	2	3	4	5	6	7	8	9	10	11	12	13	14	15
RI	0	0	0.58	0.9	1.12	1.24	1.32	1.44	1.45	1.49	1.51	1.48	1.56	1.57	1.58

3. Case Design

The objective of this study is to build an evaluation index for the design of a pillbox for patients with chronic diseases. The first step is to classify and determine the items to be scored for product design. In order to identify the relevance of the preliminary evaluation indices, semi-structured interviews were conducted. The researchers surveyed professionals in related fields for their opinions, aiming at collecting an organized evaluation indicator based on their capacity to make professional judgments, assessing the importance of each question, and applying corrections or supplements to these evaluation indices. In addition to the drafted evaluation indices, the interview contents and questions were prepared so that the interviewees were asked questions such as "Any other factors to be included as the evaluation indices?" and "What are the criteria for a pillbox for patients with chronic diseases?". The interview procedure included preparing for the interview, drafting interview contents and an outline, determining the interviewees, questioning and making records, ending the interview, and collecting and analyzing the data. After interviewing the experts, the appraisal process for a pillbox for patients with chronic diseases could be determined. In addition, assessing the professional literacy of the interviewees, the conditions of the pillbox are among the critical factors during the evaluation process. Therefore, the evaluation indices can be determined, and include the first level, i.e., the goal level, which is the eventual goal of appraising product designs. The second level is the objective level, which includes five constituent elements: functionality, structure, aesthetics, creativeness, and economy. The third level covers 19 evaluation criteria. In order to carry out systematic evaluation and analysis, the evaluation factors and constituent elements were chosen to build the hierarchical structure, as shown in Figure 1. Since numerous factors could affect the selection of interviewees, quantification of the degree of influence of each factor had to be carried out. After that, the degrees of influence of various factors were combined and calculated by utilizing a systematic approach in order to obtain quantization results. The definitions of the assessment items are as follows.

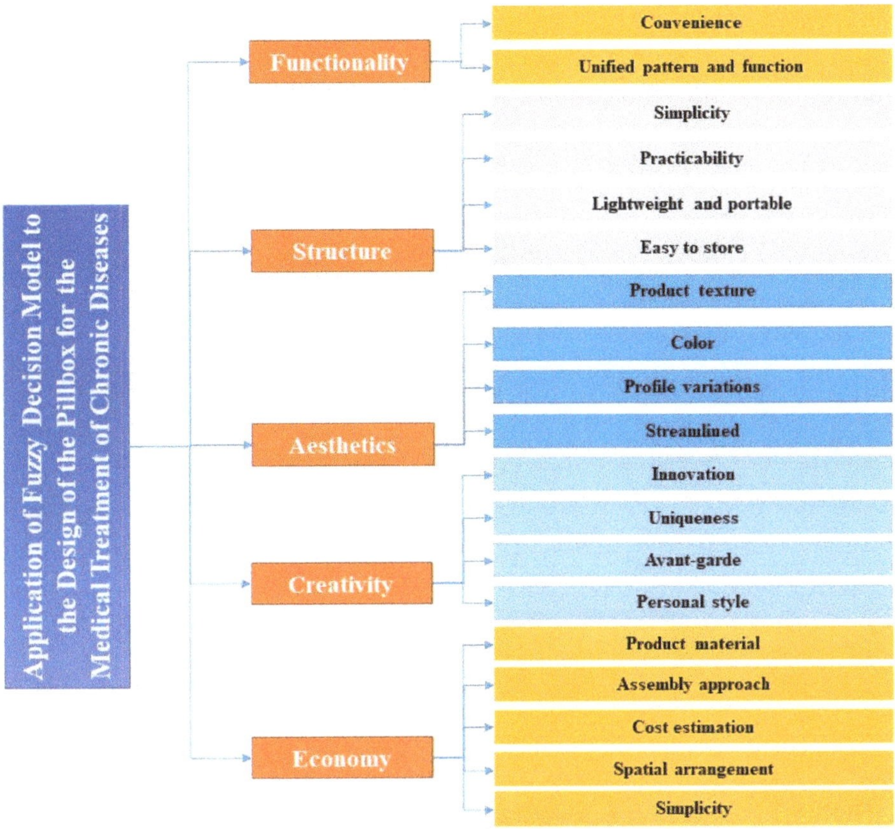

Figure 1. Hierarchical framework of evaluation indices of a pillbox for patients with chronic diseases.

First, the functionality part indicates the functionality of the pillbox for patients with chronic diseases. It covers two criteria, a product's "convenience" and its "unified pattern and function." Structure comprises "simplicity, "practicability,", "lightweight and portable," and "easy to store." Aesthetics comprises "product texture," "color," "shape variation," and "streamlined." Creativeness comprises "innovation," "uniqueness," "avant-garde," and "personal style." Economy comprises "product material," "assembly approach," "cost estimation," "spatial arrangement," and "simplicity."

In addition, the interviewees were mainly experts who could provide answers to the questions. Through the investigation and questionnaire, the resulting data were processed by the quantization approach. A total of 24 experts participated in this study, including 14 who run a business in an industry related to the pillbox and 10 workers in this industry. The information on their backgrounds is shown in Table 1. The participants were asked to consider the evaluation factors of the design of a pillbox for patients with chronic diseases. Answers, including the implementation of the AHP method and the fuzzy comprehensive evaluation, were checked. The goal was to obtain objective results from the questionnaire survey. After the relative weight of each factor was obtained by the AHP method and the consistency examination was carried out, Model 4 of the fuzzy synthetic method was used to apply the evaluation indices to obtain the final evaluation results.

Fuzzy multicriteria decision-making was used in this study to determine the evaluation indices of a pillbox design for patients with chronic diseases. Based on the above-mentioned analysis and construction of the evaluation factors for the pillbox, a total of 5 constituent elements and 19 evaluation criteria were obtained. They were classified into upper-level and lower-level factors as follows:

Upper-level factors: U = {Functionality ($U1$), Structure ($U2$), Aesthetics ($U3$), Creativity (U_4), Economy (U_5)}

Lower-level factors: U_1 = {Convenience (u_{11}), Unified pattern and function (u_{12})}

U_2 = {Simplicity (u_{21}), Practicability (u_{22}), Lightweight and portable (u_{23}), Easy to store (u_{24})}

U_3 = {Product texture (u_{31}), Color (u_{32}), Shape variation (u_{33}), Streamlined (u_{34})}

U_4 = {Innovation (u_{41}), Uniqueness (u_{42}), Avant-garde (u_{43}), Personal style (u_{44})}

U_5 = {Product material (u_{51}), Assembly approach (u_{52}), Cost estimation (u_{53}), Spatial arrangement (u_{54}), Simplicity (u_{54})}

In order to reflect the importance of each factor, the AHP was implemented to obtain the relative weights determined by the experts in this study. The questionnaire contents included a letter of instruction, instructions for filling out the questionnaire and examples, the standards of the intensity of importance, the hierarchical framework of indices, explanations, and questions. The importance of two factors in each subsystem was compared. The evaluation scale basically can be divided into five levels: equally important, slightly important, considerably important, very important, and absolutely important. They were assigned weights of 1, 3, 5, 7, and 9. An additional four levels between these five levels were assigned weights of 2, 4, 6, and 8. Left-aligned scales indicated that the factors on the left were more important than the factors on the right. On the other hand, right-aligned scales indicated that the factors on the right were more important than the factors on the left. The experts were asked to tick adequate assessment items, and the results are shown in Tables 3 and 4.

Table 3. Weight set of upper-level factors of evaluation indices for a pillbox design for patients with chronic diseases.

Ranking	Upper-level Factor	Weight	*CR*
1	Aesthetics (U_3)	0.40	
2	Structure (U_2)	0.29	
3	Creativity (U_4)	0.15	0.01
4	Functionality (U_1)	0.10	
5	Economy (U_5)	0.07	

Table 4. Weight set of lower-level factors of evaluation indices for pillbox design for patients with chronic diseases.

Upper-level Factor	Lower-level Factor	Weight	*CR*
Functionality (U_1)	Convenience (u_{11})	0.532	0.01
	Unified pattern and function (u_{12})	0.468	
Structure (U_2)	Simplicity (u_{21})	0.424	0.03
	Practicability (u_{22})	0.232	
	Lightweight and portable (u_{23})	0.312	
	Easy to store (u_{24})	0.032	
Aesthetics (U_3)	Product texture (u_{31})	0.153	0.03
	Color (u_{32})	0.241	
	Shape variation (u_{33})	0.194	
	Streamlined (u_{34})	0.412	

Table 4. *Cont.*

Upper-level Factor	Lower-level Factor	Weight	*CR*
Creativity (U_4)	Innovation (u_{41})	0.22	0.02
	Uniqueness (u_{42})	0.229	
	Avant-garde (u_{43})	0.352	
	Personal style (u_{44})	0.199	
Economy (U_5)	Product material (u_{51})	0.164	0.01
	Assembly approach (u_{52})	0.201	
	Cost estimation (u_{53})	0.167	
	Spatial arrangement (u_{54})	0.178	
	Simplicity (u_{55})	0.29	

In addition, it is known from the AHP method that when $C.R. \leq 0.1$, it can be determined that the judgment matrix has satisfactory consistency. It also demonstrated that the weight distribution is reasonable, and the results are shown in Tables 3 and 4.

Based on the results in Tables 2 and 3 and from Equation (2), the weight set of each factor can be obtained as follows:

$\widetilde{W}_1 = [0.532, 0.468]$, $\widetilde{W}_2 = [0.424, 0.232, 0.312, 0.032]$,
$\widetilde{W}_3 = [0.153, 0.241, 0.194, 0.412]$, $\widetilde{W}_4 = [0.22, 0.229, 0.352, 0.199]$,
$\widetilde{W}_5 = [0.164, 0.201, 0.167, 0.178, 0.29]$, $\widetilde{W} = [0.135, 0.286, 0.297, 0.159, 0.123]$.

Moreover, the set of evaluation results obtained from the participants' judgment of the evaluation target can be classified into five levels: V = {completely agree, agree, neither agree nor disagree, disagree, completely disagree}. Based on the factor set and evaluation set, the questionnaire for comprehensive evaluation of the designs allowed experts to fill in their answers to evaluate each factor. After the researchers collected the questionnaires for further statistical analysis, the membership grade of the evaluation of each factor was determined in order to obtain their fuzzy evaluation matrices. The fuzzy set is summarized as follows:

$$\widetilde{R}_1 = \begin{bmatrix} 0.32 & 0.48 & 0.16 & 0.04 & 0.00 \\ 0.19 & 0.29 & 0.48 & 0.05 & 0.00 \end{bmatrix},$$

$$\widetilde{R}_2 = \begin{bmatrix} 0.33 & 0.38 & 0.25 & 0.04 & 0.00 \\ 0.20 & 0.30 & 0.40 & 0.10 & 0.00 \\ 0.46 & 0.23 & 0.31 & 0.00 & 0.00 \\ 0.35 & 0.26 & 0.35 & 0.04 & 0.00 \end{bmatrix},$$

$$\widetilde{R}_3 = \begin{bmatrix} 0.20 & 0.30 & 0.40 & 0.10 & 0.00 \\ 0.21 & 0.16 & 0.53 & 0.11 & 0.00 \\ 0.21 & 0.32 & 0.32 & 0.16 & 0.00 \\ 0.35 & 0.39 & 0.17 & 0.09 & 0.00 \end{bmatrix},$$

$$\widetilde{R}_4 = \begin{bmatrix} 0.18 & 0.41 & 0.36 & 0.05 & 0.00 \\ 0.19 & 0.43 & 0.29 & 0.10 & 0.00 \\ 0.35 & 0.39 & 0.17 & 0.09 & 0.00 \\ 0.21 & 0.47 & 0.11 & 0.21 & 0.00 \end{bmatrix},$$

$$\widetilde{R}_5 = \begin{bmatrix} 0.33 & 0.50 & 0.08 & 0.08 & 0.00 \\ 0.18 & 0.41 & 0.36 & 0.05 & 0.00 \\ 0.33 & 0.38 & 0.25 & 0.04 & 0.00 \\ 0.21 & 0.32 & 0.32 & 0.16 & 0.00 \\ 0.38 & 0.29 & 0.19 & 0.14 & 0.00 \end{bmatrix}.$$

The fuzzy evaluation matrix can be obtained from the above-mentioned procedures. The comprehensive evaluation and operation were carried out by implementing Model 4 of the fuzzy synthetic method. This synthetic method only requires carrying out normalization of w_i. Therefore, there is no need to normalize the results obtained from the fuzzy comprehensive evaluation. The operations are as follows for the low-level evaluation:

$$\text{Functionality factor}: \widetilde{B}_1 = \widetilde{W}_1 \bullet \widetilde{R}_1 = \begin{bmatrix} 0.263 & 0.397 & 0.300 & 0.044 & 0.00 \end{bmatrix},$$

$$\text{Structure factor}: \widetilde{B}_2 = \widetilde{W}_2 \bullet \widetilde{R}_2 = \begin{bmatrix} 0.341 & 0.311 & 0.307 & 0.041 & 0.00 \end{bmatrix},$$

$$\text{Aesthetics factor}: \widetilde{B}_3 = \widetilde{W}_3 \bullet \widetilde{R}_3 = \begin{bmatrix} 0.266 & 0.307 & 0.321 & 0.110 & 0.00 \end{bmatrix},$$

$$\text{Creativity factor}: \widetilde{B}_4 = \widetilde{W}_4 \bullet \widetilde{R}_4 = \begin{bmatrix} 0.248 & 0.419 & 0.227 & 0.045 & 0.107 \end{bmatrix},$$

$$\text{Economy factor}: \widetilde{B}_5 = \widetilde{W}_5 \bullet \widetilde{R}_5 = \begin{bmatrix} 0.293 & 0.369 & 0.239 & 0.099 & 0.00 \end{bmatrix}.$$

It is known from Table 2 that the weights of the high-level factors can be determined, and the high-level judgment matrix is as follows:

$$\widetilde{R}^* = \begin{bmatrix} \widetilde{B}_1 \\ \widetilde{B}_2 \\ \widetilde{B}_3 \\ \widetilde{B}_4 \\ \widetilde{B}_5 \end{bmatrix} = \begin{bmatrix} 0.263 & 0.397 & 0.300 & 0.044 & 0.00 \\ 0.341 & 0.311 & 0.307 & 0.041 & 0.00 \\ 0.266 & 0.307 & 0.321 & 0.110 & 0.00 \\ 0.248 & 0.419 & 0.227 & 0.107 & 0.00 \\ 0.293 & 0.369 & 0.239 & 0.099 & 0.00 \end{bmatrix}.$$

Therefore, the result of the high-level fuzzy comprehensive evaluation is

$$\widetilde{C} = \widetilde{W} \bullet \widetilde{R}^* = \begin{bmatrix} 0.288 & 0.346 & 0.289 & 0.080 & 0.00 \end{bmatrix}.$$

As for the processing of evaluation indices, typical methods in common use are maximum degree of membership and weighted averaging. Weighted averaging can turn vague values into definite values. This is the so-called defuzzification effect. The purpose of defuzzification is to transform the final data or results with vague properties into definite values and data. If vague values are used in the operations, the result is also a vague value. Defuzzification of vague values must be done so that they can turn into definite values with their own representativeness for the benefit of follow-up comparison and ranking operations. Therefore, by calculating with the weighted-average method in this study, the concept of a hierarchy of values is applied to the results (Kuo and Chen, 2006). Assigning V = {completely agree, agree, neither agree nor disagree, disagree, completely disagree} = {1, 0.75, 0.50, 0.25, 0}, the researchers calculated defuzzified values of evaluation results D, shown in Tables 5 and 6.

Table 5. Degrees of conformity of various factors determined by interviewees.

Evaluation Factor	Completely Agree	Agree	Neither Agree nor Disagree	Disagree	Completely Disagree	Defuzzification
Functionality	0.259	0.391	0.304	0.046	0.00	0.716
Structure	0.319	0.391	0.300	0.063	0.00	0.723
Aesthetics	0.300	0.297	0.251	0.152	0.00	0.686
Creativity	0.298	0.324	0.236	0.142	0.00	0.694
Economy	0.300	0.306	0.250	0.144	0.00	0.691

Table 6. Degrees of conformity of index framework for evaluation by interviewees.

Constituent Elements of Evaluation Indices of Pillbox Design for Patients with Chronic Diseases	Completely Agree	Agree	Neither Agree nor Disagree	Disagree	Completely Disagree	Defuzzification
	0.288	0.346	0.289	0.080	0.00	0.711

In this study, the evaluation of pillbox designs for patients with chronic diseases was carried out by the fuzzy theory. The research results are shown in Tables 5 and 6. The degrees of conformity of various factors determined by the interviewees are shown in Table 5. The degrees of conformity of the index framework for evaluation by the interviewees are shown in Table 6. If the principle of maximum membership serves for the processing of evaluation indices, for the Structure factor, the evaluation items include "simplicity" (u_{21}), "practicability" (u_{22}), "lightweight and portable" (u_{23}), and "easy to store" (u_{24}). The evaluation result indicates that using the factors as constituent elements of the pillbox design is at the level

If the values obtained after defuzzification serve as evaluation indices for processing, it is known from the comprehensive evaluation that the result of the Functionality factor is 0.716, which indicates a level of "Completely agree" for the pillbox design, followed by Structure at 0.713, Aesthetics at 0.686, Creativity at 0.694, and Economy at 0.691. Since the design of a pillbox for patients with chronic diseases should follow regulated dimensions and patterns, according to the AHP, the factor weights are as shown in Tables 3 and 4. The weight of the Structure factor is 0.319, which corresponds to the ranking from the fuzzy comprehensive evaluation. However, the pairwise comparison matrix in the AHP has the problems of subjectivity, inaccuracy, and vagueness. In order to resolve this problem, the AHP approach must be extended to the vague environment in order to compensate for the deficiency of the vagueness problems that the AHP cannot resolve. After that, the fuzzy comprehensive evaluation is implemented to select the evaluation items in order to obtain their fuzzy values. Therefore, the result can be viewed as a dual verification, which indicates a certain degree of commonality between two factors in order to enhance the accuracy of the research. Moreover, the resulting values in Table 5 indicate that the overall index of the evaluation of the pillbox design is 0.723, which is between the levels "Completely agree" and "Agree." This indicates that the framework of the evaluation indices is acceptable. The result serves as a good reference for the process of designing a pillbox for patients with chronic diseases, as shown in Figure 2.

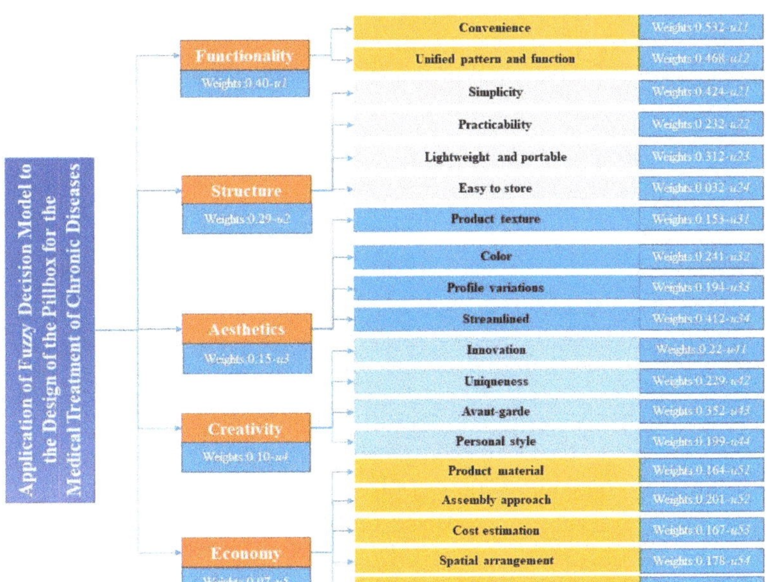

Figure 2. Weight set of the lower factors of the trolley selection indicator.

3.1. Design Case Comparison

Finally, three pillbox designs with different forms and styles are shown in Table 7. Case 1 presents a timed pillbox design as the main body and its contents show a simple and convenient design. This design provides a simple and convenient way of reminding users of the time and function of taking a pill. Case 2 presents a concept of a pillbox composed of a button-type storage compartment. This design is simple and easy to use. Case 3 features a rotating pillbox. The main idea is to implement a rotating shape in an attempt to increase the storage space. Each of these three designs has a different style.

For the evaluation of the overall schemes, it is known from Figure 3 that the defuzzified values of Case 3 is 0.74, which is at the "Satisfied" level, followed by Case 1, which has a defuzzified value of 0.62, at the "Neither satisfied nor dissatisfied" level. Therefore, Case 1 is one the most favored by subjects who made the decision. This scheme is not only full of functionality and economical, but it also performs well considering the coordination between humans, machines, and the environment. The advantage of this scheme lies in its creativity and aesthetics. Follow-up studies are advised to highlight the consideration of aesthetics. The results also indicate that Case 2 is inferior in terms of aesthetics. On the other hand, the economic aspect of Case 1 should be further enhanced as well.

Table 7. Three types of new pillbox designs with different shapes for patients with chronic diseases.

	Case 1	Case 2	Case 3
Design case			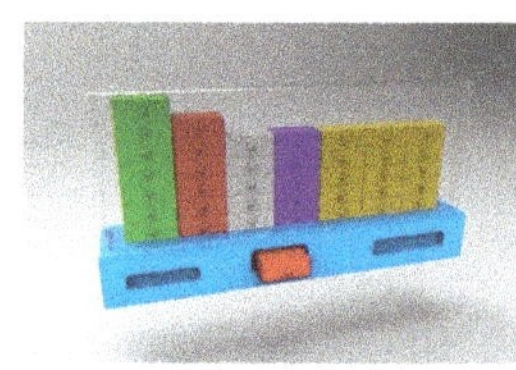
Design name	Modularization and barcode applications	Button-type pillbox	Smart barcode application with rotatable pillbox
Design case description	Application of barcodes. Each pillbox package has a QR code. Before taking a pill, the user scans the QR code and is made aware of the contents, including side effects and directions for use. The container can link to the pillbox. The weight of each pill is shown on the front panel. Users know the weight of the pillbox and the remaining pills so they know whether they forgot to take a pill and on which day.	Composed in the shape of a button-type storage compartment. The main design is simple and easy to use.	Each kit has a QR code. Before taking the pill, the user scans the QR code and knows the drug content, including side effects and instructions for use. The container can be linked to the kit. The addition of color and LED display provides the correct medication for the user, and the weight of each pill is displayed on the front panel. Users know the weight of the pillbox and the remaining pills so they know if they forgot to take the medicine and on which day.

The pillbox design of Case 3 provides the function of a favorable performance design with portable pill cells. Since patients are not professional medical personnel (Figure 4), they might take the wrong medicine by mistake or by misunderstanding the information. In order to create an intelligent pillbox that can recognize prescriptions and scheduling, a new type of portable cell is created. It is similar to the design of hospital medicine bags that can prevent cross-contamination of different medicines. Each cell is loaded with only one type of medicine and there is a barcode for the medicine information on the top. The information includes patient name, medicine name, and dose, as well as medication instructions. Based on doctors' prescriptions, pharmacists load the medicines into the portable cells and seal them for patients.

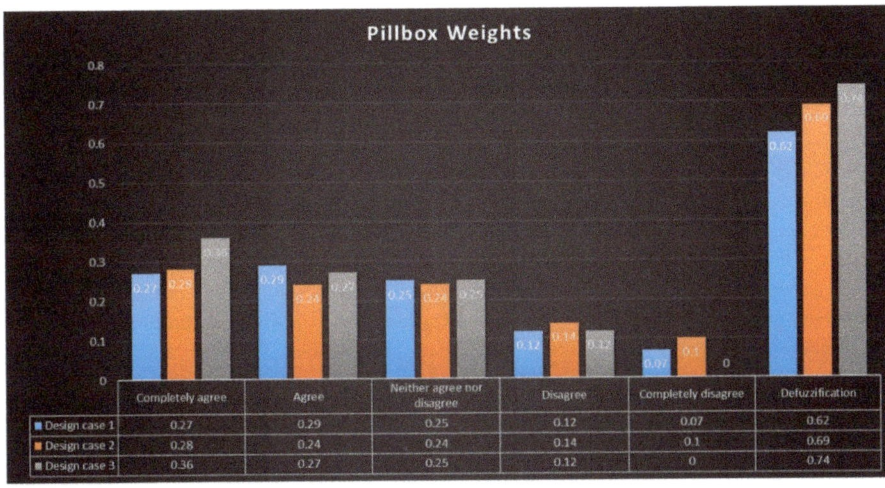

Figure 3. Subjects' satisfaction degree and defuzzification values for the three design cases. satisfaction degree value, the highest score of each program.

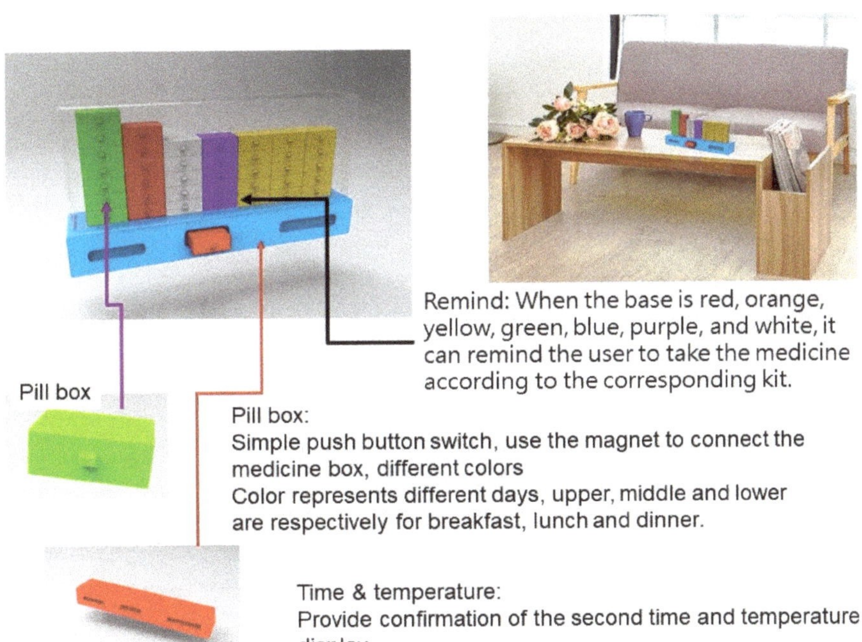

Figure 4. Pillbox control module mock-up.

In addition, the design of the pillbox control module is to house the portable medicine cells. The top cover is opened to display the barcodes of the cells so that patients can scan for prescriptions. Moreover, in order to effectively avoid medication errors, the pillbox control module is equipped with a mechanical device that includes an electronic lock, sensors, and LEDs for controlling and managing the cells. The functions of the control module include detecting whether the portable medicine cell is loaded and the top cover is sealed, locking the pillbox, and indicating medicine locations by LEDs, etc.

4. Conclusions

From the case study and verification of its applications, the results in Table 2 indicate the design of a pillbox for patients with chronic diseases, which is part of the aesthetics study of the entire evaluation of items. Not only were some of the designs chosen, but their aesthetics were improved. Finally, some satisfactory designs were determined by the participants. The score of Case 3 after defuzzification is 0.74, which is at the satisfactory level, followed by 0.69 for Case 2, which is at the common level. Therefore, the overall evaluation result is favored by the decision-making group in Item 1. This design scheme not only provides functionality, aesthetics, and economy, but the functions of the pillbox design for patients with chronic diseases show favorable performance. That is to say, the procedure is to determine the optimal design that can be realized. In this study, a new AHP method was proposed, featuring the integration of fuzzy theory and numerical analysis. First, the AHP is used to calculate the weight of factors affecting the importance of performance parameters. The next step is to satisfy various parameters. The defuzzification procedure is implemented to obtain the results of fuzzy weights, which indicated good consistency of the numerical results. The fuzzy decision-making method can realize any engineering evaluation that is required for the overall product design of a pillbox for patients with chronic diseases. The emphasis of this method is on the overall evaluation of the design and the new performance evaluation. In other words, not only the design of the product set of chronic medicine, but also the entire evaluation method is improved. The results can serve as a basis for an overall evaluation model of the product set for chronic medicines based on the new aesthetic design.

Funding: This work was supported by the Ministry of Science and Technology of the Republic of China under grant MOST-108-2221-E-468-003.

Conflicts of Interest: The authors declare no conflict of interest.

References

1. Menachem, P.W. Amihud Hari, Extension of the Pahl & Beitz systematic method for conceptual design of a new product. *Procedia CIRP* **2015**, *36*, 254–260.
2. Gielisch, C.; Fritz, K.P.; Noack, A.; Zimmermann, A. A Product Development Approach in The Field of Micro-Assembly with Emphasis on Conceptual Design. *Appl. Sci.* **2019**, *9*, 1920. [CrossRef]
3. Fiorineschi, L.; Frillici, F.S.; Rotini, F.; Tomassini, M. Exploiting TRIZ Tools for enhancing systematic conceptual design activities. *J. Eng. Des.* **2018**, *29*, 259–290. [CrossRef]
4. Becattinia, N.; Cascinia, G.; Rotinib, F. OTSM-TRIZ Network of Problems for evaluating the design skills. *Procedia Eng.* **2015**, *131*, 689–700. [CrossRef]
5. Fiorineschi, L.; Frillici, F.S.; Rotini, F. Three-dimensional approach for assessing uncommonness of ideas. *Proc. Des. Soc. Int. Conf. Eng. Des.* **2019**, *1*, 229–238. [CrossRef]
6. Jenab, K.; Sarfaraz, A.; Ameli, M.T. A conceptual design selection model considering. *J. Eng. Des.* **2013**, *24*, 293–304. [CrossRef]
7. Erden, M.S.; Komoto, H.; van Beek, T.J.; D'Amelio, V.; Echavarria, E.; Tomiyama, T. A review of function modeling: Approaches and applications. *Articial Intell. Eng. Des. Anal. Manuf.* **2008**, *22*, 147–169. [CrossRef]
8. Eckert, C. That which is not form: The practical challenges in using functional concepts in design. *Articial Intell. Eng. Des. Anal. Manuf.* **2013**, *27*, 217–231. [CrossRef]
9. Ulrich, K.T.; Eppinger, S.D. *Product Design and Development*, 5th ed.; Mc Graw Hill Irwin: New York, NY, USA, 2012.
10. Toh, C.A.; Miller, S.R. How engineering teams select design concepts: A view through the lens of creativity. *Des. Studies.* **2015**, *38*, 111–138. [CrossRef]
11. Ng, C.Y. Evidential reasoning-based Fuzzy AHP approach for the evaluation ofdesign alternatives' environmental performances. *Appl. Soft Comput.* **2016**, *46*, 381–397. [CrossRef]
12. Lin, H.H.; Hsiao, S.W. A Study of the Evaluation of Products by Industrial Design Students. *EURASIA J. Math. Sci. Technol. Educ.* **2018**, *14*, 239–254. [CrossRef]
13. Lin, M.C.; Wang, C.C.; Chen, M.S.; Chang, C.A. Using AHP and TOPSIS approaches in customer-driven product design process. *Comput. Ind.* **2008**, *59*, 17–31. [CrossRef]

14. Hsiao, S.W. Fuzzy logic based decision model for product design. *Int. J. Ind. Ergon.* **1998**, *21*, 103–116. [CrossRef]
15. Moskowitz, H.; Kim, K.J. QFD optimizer: A novice friendly quality function deployment decision support system for optimizing product designs. *Comput. Ind. Eng.* **1997**, *32*, 641–655. [CrossRef]
16. Jin, Y.; Cao, W.; Wu, M.; Yuan, Y. Accurate fuzzy predictive models through complexity reduction based on decision of needed fuzzy rules. *Neurocomputing* **2019**, *323*, 344–351. [CrossRef]
17. Kuo, Y.F.; Chen, P.C. Selection of mobile value—Added services for system operators using fuzzy synthetic evaluation. *Expert Syst. Appl.* **2006**, *30*, 612–620. [CrossRef]
18. Sun, J.H.; Tsai, H.C.; Chen, S.I. Using triangles to blur the preferences of different types of consumers. In Proceedings of the 17th National Conference on Fuzzy Theory and Its Applications, Kaohsiung, Taiwan, 18–19 December 2009; pp. 396–401.
19. Hu, T.W.; Liao, Y.T. The Effects of Consumer's Visual Cognition on the Feature Composition of Bicycle Frames. *J. Chaoyang Univ. Technol.* **2010**, *15*, 283–300.
20. Hajek, P.; Froelich, W. Integrating TOPSIS with interval-valued intuitionistic fuzzy cognitive maps for effective group decision making. *Inf. Sci.* **2019**, *485*, 394–412. [CrossRef]
21. Zadeh, L.A. Fuzzy sets. *Inf. Control* **1965**, *8*, 338–353. [CrossRef]
22. Zadeh, L.A. The concept of a linguistic variable and its application to approximate reasoning I. *Inf. Sci.* **1975**, *8*, 199–249. [CrossRef]
23. Depeng, K.; Tianqing, C.; Jiali, P.; Na, H.; GuoZhen, Y. A decision variable-based combinatorial optimization approach for interval-valued intuitionistic fuzzy MAGDM. *Inf. Sci.* **2019**, *484*, 197–218.
24. Nazemi, A.; Pour, F.F.; Heidenreich, K.; Fabozzi, F.J. Fuzzy decision fusion approach for loss-given-default modelling. *Eur. J. Oper. Res.* **2017**, *262*, 780–791. [CrossRef]
25. Temponi, C.; Yen, J.; Tian, W.A. House of quality: A fuzzy logic—Based requirements analysis. *Eur. J. Oper. Res.* **1999**, *117*, 340–354. [CrossRef]
26. Fang, X.; Yang, Q.; Dong, W. Fuzzy decision based energy dispatch in offshore industrial microgrid with desalination process and multi-type DGs. *Energy* **2018**, *148*, 744–755. [CrossRef]
27. Kim, J.K.; Han, C.H.; Choi, S.H. A knowledge-base approach to the quality function deployment. *Comput. Ind. Eng.* **1998**, *35*, 233–236. [CrossRef]
28. Saaty, T.L. *The Analytic Hierarchy Process*; McGraw-Hill New York: New York, NY, USA, 1980.
29. Hsiao, S.W.; Tsai, H.C. Fuzzy set theory on car-color design. *Color Res. Appl.* **1994**, *19*, 202–213. [CrossRef]
30. Hsiao, S.W. A systematic method for color planning in product design. *Color Res. Appl.* **1995**, *20*, 191–205. [CrossRef]
31. Hsiao, S.W.; Chen, C.H. A semantic and shape grammar based approach for product design. *Des. Stud.* **1997**, *18*, 275–296. [CrossRef]
32. Hsiao, S.W.; Chang, M.S. A semantic recognition-based approach for car's concept design. *Int. J. Veh. Des.* **1997**, *18*, 53–82.
33. Hsiao, S.W. Fuzzy set theory applied to car style design. *Int. J. Veh. Des.* **1994**, *15*, 255–278.
34. Hsiao, S.W.; Fan, C.W. Integrated FSM, STM and DFA method to faucet design. *J. Chin. Inst. Ind. Eng.* **1996**, *13*, 225–235.
35. Lin, H.H. A systematic approach of predicting color preference on the basis of gray relational grade. *Color Res. Appl.* **2019**, *44*, 194–204. [CrossRef]

applied sciences

MDPI

Article

Application of Fuzzy Theory to the Evaluation Model of Product Assembly Design and Usability Operation Complexity

Ya-Chuan Ko [1,2]

[1] Department of Creative Product Design, Asia University, Taichung 41354, Taiwan; chrisko@asia.edu.tw; Tel.: +886-4-2332-3456

[2] Department of Medical Research, China Medical University Hospital, China Medical University, Taichung 40445, Taiwan

Received: 21 August 2019; Accepted: 23 September 2019; Published: 27 September 2019

Abstract: In order to make people's lives more convenient and enhance living standards, the composition of a product usually includes more than one component. However, a product is created by the joint endeavor of people from various territories and therefore one of the important considerations is making a product merge into consumers' daily lives rather than simply fulfilling its functions. There might be conflicts with people's existing life patterns or existing values, which should be taken into consideration during the manufacturing process. This study is an investigation of the process of assembly by considering the assembly operations and the assembly operating time. By determining the relationship between components, the assembly concept of most components was analyzed. A fuzzy comprehensive evaluation was carried out during the evaluation of the degree of complexity of user operations. Depending on the ranking of membership, the most appropriate assembly was determined and this serves as a reference for designers to select the optimal product assembly. By recording the consumer usage models, the optimal assembly and usage model of product design were also proposed. The goal of this study is to find the balance between assembly evaluation and user usage model by the process in order to allow designers to determine the new assembly concepts that meet consumer usage models. A case study of four bedside stereos was carried out by implementing the proposed approach in order to determine the evaluation principle of assembly. The purpose of this is to enhance the balance between assembly design and user operation complexity for making efficient decisions. A product design can comply with the spirit of concurrent engineering and the quality of a product design can be enhanced.

Keywords: assembly design; usability and operation complexity; fuzzy theory

---

1. Introduction

Nowadays, consumer preferences are highly changeable, under a limited amount of resources, and it is difficult for a company to meet the demands of a variety of consumers of the stereo market. Being confronted with this condition, a company needs to consider a way of determining the most beneficial market for itself and to serve the consumers of that market in an efficient way. Due to the fact that typical design evaluation methods are not capable of correcting the evaluation of fuzzy messages from the aesthetic aspect, a new approach of assisting consumers in selecting stereo speakers by implementing the fuzzy theory has been developed. Lo et al. [1] proposed the use of the fuzzy breaking-circle method on the design of product interface operation and layout. Ko et al. [2] and Ko [3] proposed the analysis of the hot zones of users' product operability to carry out the re-arrangement of layout design for analysis.

Quantitative approaches were proposed by scholars such as Kwahk and Han [4], Lee et al. [5], Lee et al. [6], Heo et al. [7] and Jin and Ji [8] for carrying out the principles of usability evaluation respectively. In their studies, they introduced the usability risk level evaluation during the earlier stage of conceptual designs. This is due to the fact that a design at the earlier stage might affect the usability interface. Moreover, from the aspect of product designs, the decomposition and analysis of a product by perceptions and practical operations was proposed by Chou [9], Hsiao et al. [10] and Ko et al. [2] in order to determine the optimal solution.

Therefore, the goal of this study is to determine the appropriate design case study that can demonstrate the development of this type of research method. Considering the expected product attributes, the product to be reviewed in this study should be at the mature stage of its life cycle so that the design consideration focuses on the manufacturing and assembly stages. Stereo speakers serve as a good case study as they are mostly assembled by screws or latches in a similar way to other home appliances. The main approach is to use the morphological charts to classify various components in order to simplify a problem. An experiment was also carried out by showing the participants mechanical simulation videos of a product. The design ideas were presented in the form of 3D models by excluding the consideration of color and material. The aim of this study is to achieve the following goals:

- By analyzing the theories that are relevant to this study, new concepts of stereo speakers can be generated along with 3D models that can simulate mechanical operations.
- By analyzing the theories that are relevant to this study, quantized values of the psychological perception of consumers can be obtained.
- By analyzing the theories that are relevant to this study, new design ideas can be ranked according to their assembly efficiency and assembly cost.
- By analyzing the theories that are relevant to this study, new design ideas that comply with a company's decisions from a variety of aspects can be determined.
- By analyzing assembly approaches and evaluating the usability, design ideas that have a higher value of development can be determined.
- By analyzing assembly approaches and evaluating the usability, the application of concurrent engineering can be enhanced.

2. Materials and Methods

2.1. Introduction to Assembly Design

In this section, the literature that is relevant to the assembly process of components of a product and its evaluation is reviewed.

2.1.1. Planning of the Assembly Process

In order to carry out systematic planning and analysis of product assembly, various approaches of assembly planning and determining the ranking and principles of evaluations have been proposed by researchers. Depending on their methods or aspects, they are described respectively as follows:

- Review of direct-assembly approach: This approach deals with the serial assembly of a product starting from the first component. A product is complete after assembling the last component. This type of inference method is also called the goal driven or backward chaining approach.
- Review of reverse-assembly approach: This approach was investigated from the standpoint that if the disassembly process of a product is appropriate, the adequate assembly sequence of a product can be derived from the reverse sequence of the disassembly sequence.

In addition, there are also reviews of the approach of decomposing and depicting the structural diagram and the approach that uses matrices for analysis.

Barbosa and Carvalho [11] proposed a guidance of applying the concept of design for manufacturing and assembly (DFMA) to specific applications such as airplane design and manufacturing. Via the traditional design for assembly approach, Moultrie and Maier [12] proposed a new approach to encompass an original assessment tool and its delivery process. Their approach included custom designed post-it notes and simple check lists for scoring. Tiwari et al. [13] integrated design for manufacturing (DFM) and design for assembly (DFA) into information communication technologies (ICT) to establish a design for manufacturing and assembly/disassembly with the joint design of products and production systems in order to realize the target standard of Industry 4.0. Sossou et al. [14] proposed a process-independent, structured and systematic method for designing assembly-free mechanisms.

2.1.2. Method of Evaluating Assembly

The relationship between a product's design and its assembly approach affects its cost. Approaches that have been used to evaluate assembly design in various industries are summarized as follows:

- Hitachi's assembly evaluation method.
- Lucas' method of analyzing assembly designs.
- Fujitsu's capacity evaluation system.
- Boothroyd Dewhurst's method of evaluating assembly designs [15].

2.2. *Usability Engineering*

In recent years, usability engineering has become one of the most important measurement indices of various product and interface designs. Norman [16], a cognitive psychologist, proposed the user-oriented thinking in order to see the whole picture of "human/machine/environment" and "human/product/scenario." He proposed an approach that allowed users to accept a product in a natural way by following his/her own mental characteristics without forcing the users to build a new mental model as shown in Figure 1.

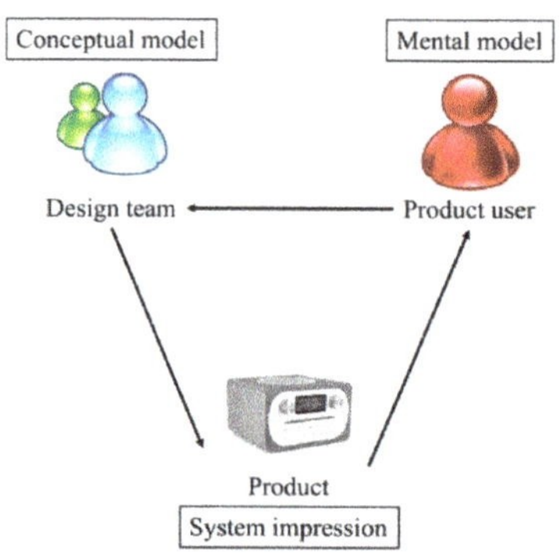

Figure 1. Mental model of a stereo system (adapted from Norman [16]).

2.3. *Fuzzy Theory*

The fuzzy theory bridges the gap between the precise classical mathematics and the real world that is full of vagueness. When encountering a problem with a high degree of vagueness and uncertainty, the fuzzy set of membership function can be used to quantize the perception of designers and consumers, as it is known that, differentiation can be realized only by comparison. During production, scientific research and our daily life, people always compare things and evaluate their quality in order to determine the optimal way of handling them. These activities include the evaluation of a member's quality, the quality of goods and the evaluation of how reasonable certain engineering design parameters are. Since a thing/object has various attributes and is affected by various factors, carrying out comprehensive consideration and fuzzy factors on several relevant factors during the process of evaluating things/objects is required. This process is called the fuzzy comprehensive evaluation.

2.4. Product Assembly Design Case

In order to find the suitable and applicable design cases and to describe the development of research methodology in a clearer way, the positioning of the product attributes in this study is on those products that are at the mature stage of product life cycle. The design consideration highlights the manufacturing and assembly orientation. Therefore, stereo systems that are available on the market were selected for the case study. Similar to other home appliances, a stereo system is assembled by screws or rivets and it is expected that the result can serve as a good reference for other relevant studies.

3. Research Process and Methods

In this study, the disassembly, ranking and evaluation approaches for assembly designs were applied to the target product. To start with the process, component drawings were processed to form the component matrices. After the ranking of assembly evaluation, the simulated scenarios of users can be developed by the usability engineering and the way how the participants were using the assembly was recorded. The fuzzy theory was implemented to the evaluation of product assembly and the degree of user preference was compared an analyzed. The purpose was to determine the correlation between the assembly consideration from the engineering orientation and the usability consideration from the user orientation. The flowchart and framework of this study is shown in Figure 2.

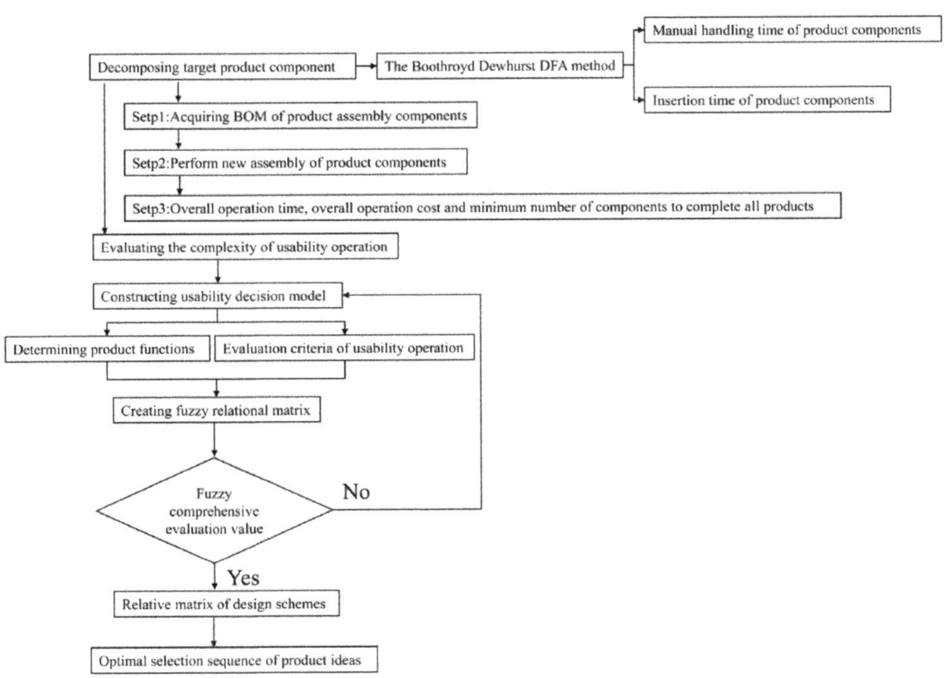

Figure 2. Flowchart and framework of this study.

3.1. Assembly Evaluation

The Boothroyd Dewhurst (DFA) method [17] was used as the main approach of evaluating assemblies. Its function is to analyze the manual processing time and installation time for components that are affected by several parameters which include component size, component weight and component thickness. The evaluation table is shown in Table 1 as follows.

Table 1. DFA evaluation table.

A	B	C	D	E	F	G	H	I	Name of Assembly
Component code	Number of continuous operations	Manual gripping code	Estimated processing time of individual gripping of component	Manual insertion code	Manual insertion time of individual component	Overall operating time (BD+F)	Operation cost	Investigating component necessity from four problems	
Totals:						Tm=	Cm=	Nm=	

The implementation procedure includes:

Step 1: Acquire the information that is related to product assembly and this information includes engineering drawings, three-dimensional exploded view drawing, existing product samples and prototypes. The information of each component is recorded in the design table.

Step 2: Perform new ways of assembling product components. The table items are ranked according to the assembly codes from the highest code to the lowest one so that a new way of assembling components can be created.

Step 3: Record the total operating time, the total operating cost and the minimum number of components after all items of product components are complete.

3.2. Fuzzy Theory

The fuzzy theory was proposed by L.A. Zadeh, a researcher and professor emeritus of computer science at the University of California, U.S. He proposed the concept of fuzzy sets in 1965 and fuzzy mathematics became a new branch of mathematics.

In order to consider the influence of all factors comprehensively so that a correct evaluation result can be obtained, the fuzzy comprehensive evaluation was proposed. By including the weights of corresponding factors, the comprehensive influence of all factors is reflected in a reasonable way. Therefore, the fuzzy comprehensive evaluation table is as follows.

$$\begin{gathered} \widetilde{B} = \widetilde{A} \cdot \widetilde{R} \\ A = (a_1, a_2, \cdots, a_m) \\ R = \begin{pmatrix} r_{11} & r_{12} & \cdots & r_{1n} \\ r_{21} & r_{22} & \cdots & r_{2n} \\ \vdots & \vdots & \vdots & \vdots \\ r_{m1} & r_{m2} & \cdots & r_{mn} \end{pmatrix} \end{gathered} \tag{1}$$

4. Results and Analysis

4.1. Selection of the Concept Schemes

In general, there are two ways of selecting the concept schemes: (1) Using the design for functions (DFF) to build a functional structure that is based on consumer demands; and (2) Using the method of morphological chart to develop design ideas that are based on the functional structure.

4.1.1. Determining Product Functions

The verb-noun approach that was proposed by Tuttle [18] among the DFF approaches was used to convert the user demands that were collected from questionnaire on stereo systems to the functional aspect of a product.

The new functions to be developed for a stereo system can be defined by adding a verb to a noun according to this approach. The results are summarized in Table 2 as follows.

Table 2. Definition of functions of a stereo system.

Verb	Noun	B.F. (Basic Function)	S.F. (Secondary Function)	A.F. (Aesthetic Function)	U.F. (Useless Function)
Play	Music	V			
Protect	Disc	V			
Avoid	Vibration	V			
Adjust	Volume	V			
Adjust	Audio source (FM)	V			
Move	Position		V		
Operating	Convenience	V			
Allow	Stack		V		
Collect	Power cord		V		
Collect	MP3		V		
Observe	Aesthetic appearance			V	
Configure	Time		V		
Configure	LCD color			V	
Reduce	Cost	V			
Reduce	Component count	V			
Assemble	High efficiency	V			
Assemble	Short time	V			
Reduce	Noise		V		
Avoid	Wrap		V		
Disassemble	Convenience		V		
Observe	Brief			V	
Observe	Elegance			V	
Observe	Brightness			V	
Perceive	Comfort			V	
Configure	Sound field		V		
Support	Different media source	V			
Position	CD player		V		
Operate	Interface		V		
Merge	Speaker		V		
Avoid	Moisture				V

4.1.2. Developing the Product Structure Tree Diagram

The product functions that are listed in Table 2 will be further described by a function structure diagram as follows. During the development stage of a concept, the most important goal of the FSD is to disassemble the main product into problems with which the details can be managed. In order to further understand product functions, a product can be defined by three abstract levels, which include functions of the whole product, functions of the subassembly and functions of the individual products.

In the FSD, basis functions are bounded by solid rectangles, secondary functions are bounded by dashed rectangles and aesthetic functions are bounded by dotted rectangles. Therefore, the FSD of a stereo system is shown in Figure 3 as follows.

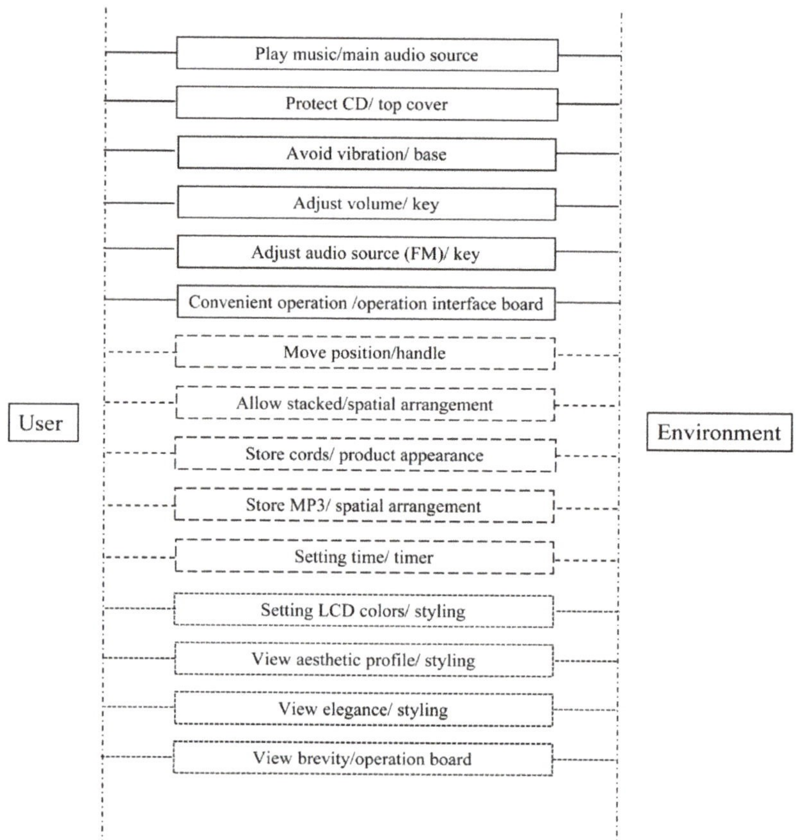

Figure 3. Function structure diagram of a stereo system.

4.1.3. Creating Concept Schemes

After the function structure diagram of a stereo system is created, a designer can further generate design ideas. The way of generating design ideas include design by analogy, brainstorming and so forth. Each approach has its own pros and cons. In this study, the design ideas were created by the morphological chart method that was proposed by Cross [19].

As shown in Table 3, after calculating the possible answer of each row, the number of all possible concepts is $4 \times 4 \times 4 \times 4 \times 4 \times 2 \times 2 \times 2 \times 2 = 16{,}384$.

Table 3. Morphological chart method.

Necessary Function	Sub-Function			
P1 Way of placing the CD	P11 Open-lid Type	P12 Slot type	P13 Suction type	P14 Slide type
P2 CD-ROM location	P21 Top	P22 Bottom	P23 Front	P24 Middle
P3 Control button style	P31 Square	P32 Round	P33 Triangular	P34 Oval
P4 LCD dimensions and size	P41 Overall	P42 3/4 area	P43 1/2 area	P44 1/4 area
P5 Control button location	P51 Top	P52 Bottom	P53 Front	P54 Edge
P6 Mp3/iPhone device	P61 Yes	P62 No		
P7 Speaker	P71 Integrated	P72 Separated		
P8 Stacked design	P81 Allowed	P82 Not allowed		
P9 Remote control	P91 Yes	P92 No		

In order to simplify the problem, the subassemblies were assembled into four concept schemes:

Idea 1. Product code: P11, P21, P32, P44, P53, P61, P72, P82, P91
Idea 2. Product code: P14, P23, P31, P41, P51, P61, P71, P82, P91
Idea 3. Product code: P12, P24, P32, P44, P52, P61, P72, P81, P91
Idea 4. Product code: P13, P24, P31, P43, P51, P61, P71, P82, P91.

4.2. Process of the Usability Experiment

Parameter models of the concept schemes that were generated in the earlier stage can be created by commercial CAD software such as Solidworks. In order to make the mechanism of those concept schemes movable, the concept schemes were disassembled into components so that the mechanism patterns can be revealed. On the other hand, the assembly efficiency and assembly time of the components were also calculated from the engineering perspective.

4.2.1. Pre-Procedure

Before creating the 3D models, the factor of colors was excluded so that these four concepts were not affected by colors. Moreover, blue dots in the videos indicate the location of a participant's hand when using the stereo systems. To distinguish the LCD monitor from a stereo system model, the area of the message interface is in dark blue colors.

The three-view drawings, specifications and the simulation of product mechanism for these four concepts are shown in Figures 4–7 and Table 4. The simulation of product mechanism includes the actions of taking out or placing the CD, adjusting volume, external source (MP3, iPhone) and so forth. The videos allow participants to understand how a concept operates and what the human-machine interaction looks like. Three-dimensional CAD software SolidWorks and the rendering and Three-dimensional rendering animation software KeyShot were used in this study.

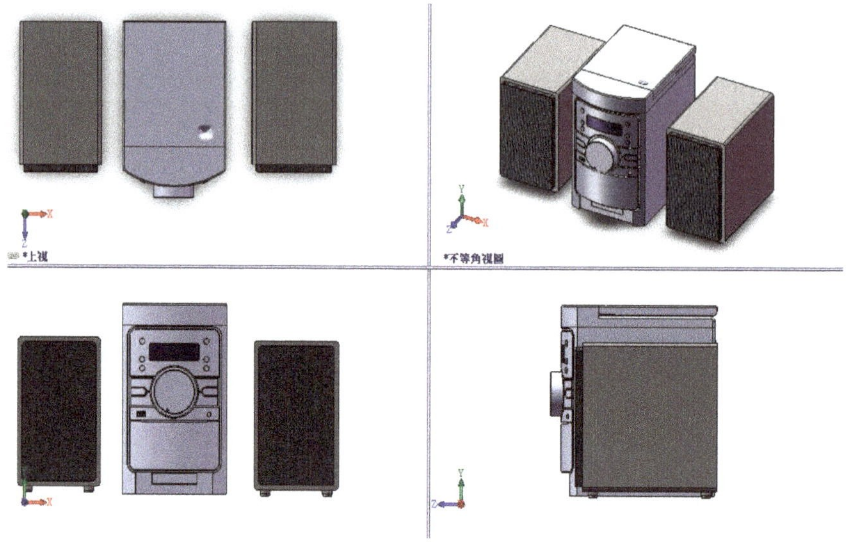

Figure 4. Three-view drawing of Idea 1.

Table 4. Product specifications of Idea 1.

Main Dimensions (mm)	140 (Width)	230 (Height)	216 (Depth)
Speaker dimensions (mm)	130	230	150
Product weight in package (kg)		4.75	
LCD dimensions (mm)	70	22	R = 2
Input interface	USB socket	Headset audio source input	iPhone
Output interface	Speaker output socket		

(a) Mechanical operation of the CD drive in a stereo system

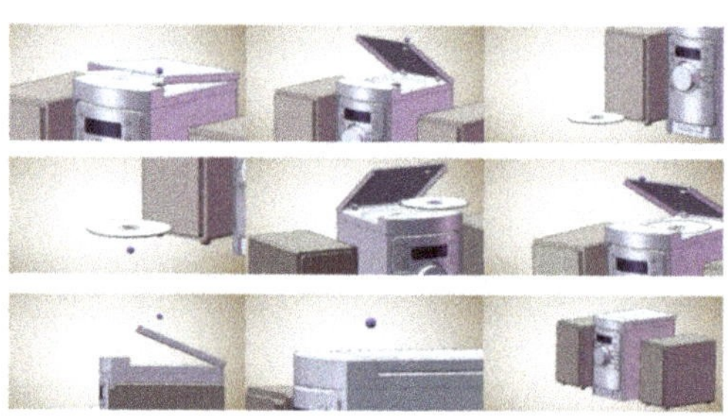

Figure 5. Animation screenshots of mechanical simulation for Idea 1.

(b) Adjusting volume

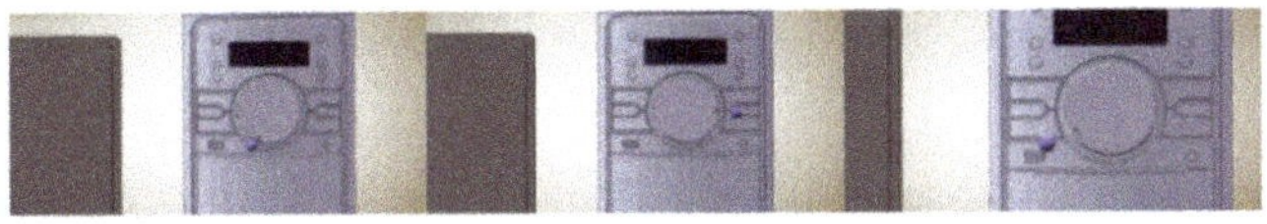

Figure 6. Animation screenshots of mechanical simulation for Idea 1.

(c) External source (iPhone/USB)

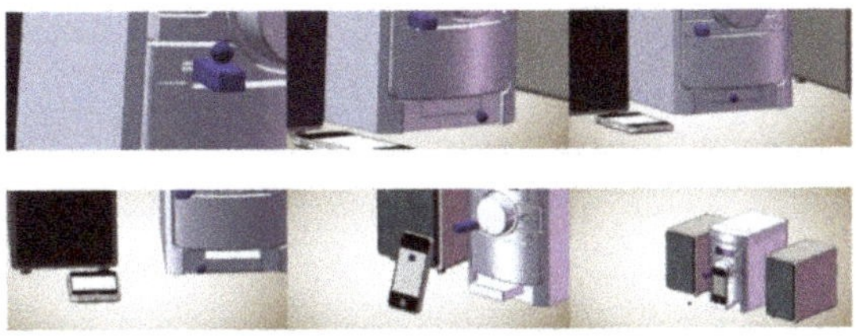

Figure 7. Animation screenshots of mechanical simulation for Idea 1.

In order to carry out the usability experiment, videos of the mechanical simulations for Idea 2, 3, 4 were also prepared. After these videos were ready, the next step was to design the questionnaire for the participants including potential customers, management and design team. The highlights of this questionnaire are as follows:

1. Learnability and efficiency
2. Memorability and errors
3. Function
4. Emotion
5. Satisfaction

4.2.2. Pre-Test

After the questionnaire and the videos of concepts were ready, two students who had design related background were invited to carry out the overall experiment prior to the formal experiment in order to check whether the experimental process was smooth and whether the questionnaire was adequate.

The main problems that were determined after this pre-test are described as follows:

6. Unclear description of some questions in the questionnaire.
7. The ranking of the preference of these design ideas was carried out before playing the videos. The researchers did not check whether the ranking changed after the end of these videos.
8. The videos were played at different speeds.
9. It is recommended using the same computer to play the videos in order to control the variables including colors under the same playing environment.

4.2.3. Experimental Procedure

The procedure of the experiment is described as follows:

10. Two laptop computers were used for the experiment. One for the participants to fill in the questionnaire and the other one for playing the simulation videos of the design ideas.
11. A CD stereo system was placed in front of the desk so that the participants can use it for the vivid simulated operations during the experiment.

12. All of the prototype stereo systems have consistent colors so that participants were not affected by colors when using the prototypes. The interface which displays the music messages are in dark blue and a participant's hand position is simulated by the blue dots.
13. After a participant viewed the three-view drawings of these four prototype products, he/she was asked to fill in his/her preference in sequence.
14. The participant was then asked to watch the mechanical simulation videos of Design Concepts 1–4 and to complete the questionnaire for these four design ideas as shown in Figures 8–11.
15. The participant was asked to fill in his/her preference of the mechanical simulation videos.

Figure 8. Experimental environment.

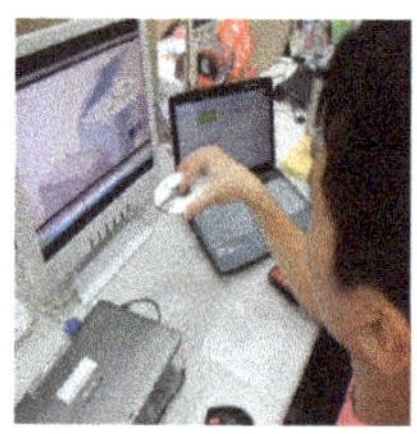

Figure 9. Simulation videos.

Figure 10. Simulation videos.

Figure 11. Simulation videos.

4.3. Results and Analysis of the Usability Experiment

A total of 30 participants were invited for the experiment in this study and their background information is shown in Table 5 as follows.

Table 5. Background information of the participants.

Variable		Count	Percentage (%)
Gender	Male	16	53.3%
	Female	14	46.7%
Age	20~30 years old	30	100%
	More than 30 years old	0	0%
Educational background	Design related	15	50%
	Not design related	15	50%

4.3.1. Outline of the Usability Decision Model

The outline of the usability decision model that was used in this study includes several steps as follows:

1. Determining the targets and the evaluation criteria and constructing the hierarchical tree diagram of objects.
2. Determining the weights by considering the relative importance of criteria.
3. After measuring the membership functions of the targets at the lowest level in the tree, building up the fuzzy relation matrix.
4. Using the weight set and the fuzzy relational matrix that was obtained from Step 2 and Step 3 to determine the fuzzy comprehensive evaluation values.
5. Selecting the optimal choices from the design schemes' relative matrices that were formed by the fuzzy comprehensive evaluation values.

4.3.2. Description of Problems

A stereo system was selected as the case study. In addition to meeting the functional, physical and styling limitations, the product is required to satisfy other limitations in the specific tasks. The first step of the evaluation is to stipulate the design guidelines. The limitations could be due to the engineering, ergonomic, aesthetic, marketing, logistic, planning, design, product and economic factors.

4.3.3. Constructing the objective tree

The method of objective tree [16,17] provides a clear format of the hierarchical relationship between higher levels and lower level during the decision-making process of design schemes. An objective tree that was constructed based on the above-mentioned design criteria is shown in Figure 12.

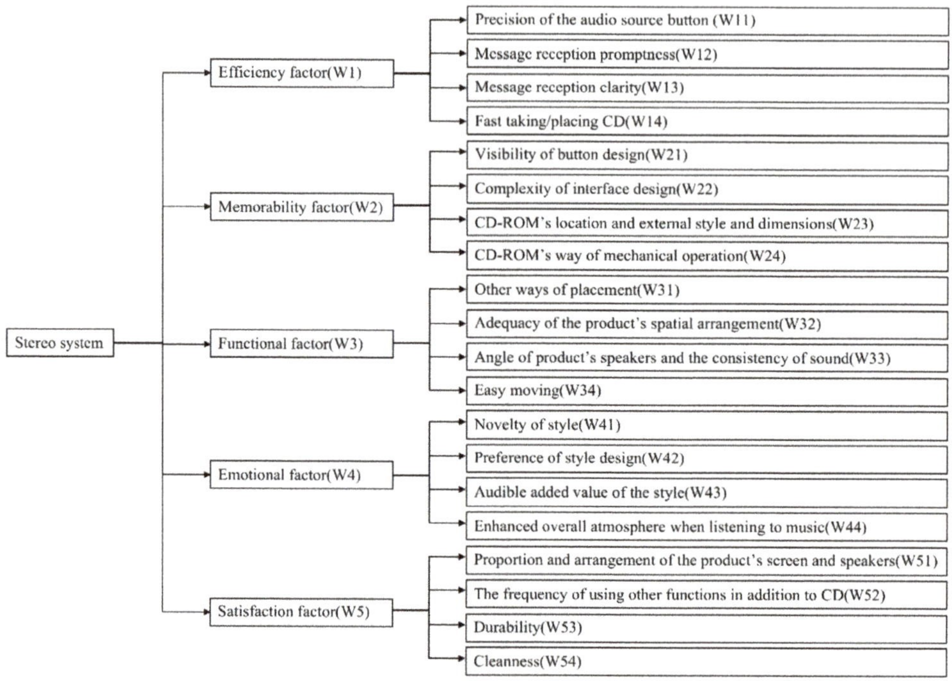

Figure 12. The objective tree of a stereo system.

4.3.4. Evaluation of weight function

In order to determine the influence of the lower levels on the higher levels, the weight functions should be determined by methods such as AHP [20], tabulated judgment method [21] and so forth. In this study, the tabulated judgment method was used for determining the weight functions.

The procedure is described as follows:

(1) Determining the relative importance of pairwise items including the F1, F2, F3 and F4 child items.
(2) List the test items into rows and columns in a table.
(3) Filling in the scores of pairwise importance into the corresponding columns according to the following principles. If two child items have the same importance as the higher level, both child items were given a score of two points. For example, if F2 is more important than F1, than F2 gets 3 points and F1 gets 1 point. If F1 is much more important than F4, F1 gets 4 points and F4 has zero point. No value is filled into the diagonal columns as this is comparison to itself.
(4) The summation of all child items *Ki* can be obtained by summing up the comparison value of each row.
(5) The summation of all child items can be aggregated into the grand total of a system with n items by the following equation.

$$\sum{}_{i=1}^{n} K_i = \left\{\left(n^2 - n\right)/2\right\} \times 4 = 2\left(n^2 - n\right) \tag{2}$$

(6) The weight function of each item can be expressed by Wi = Ki/$\sum_{i=1}^{n} K_i$, which is the weight of the nth item.

By implementing the above-mentioned method, the degree of influence of the five items and their child items can be determined as shown in Figure 13 and Table 6.

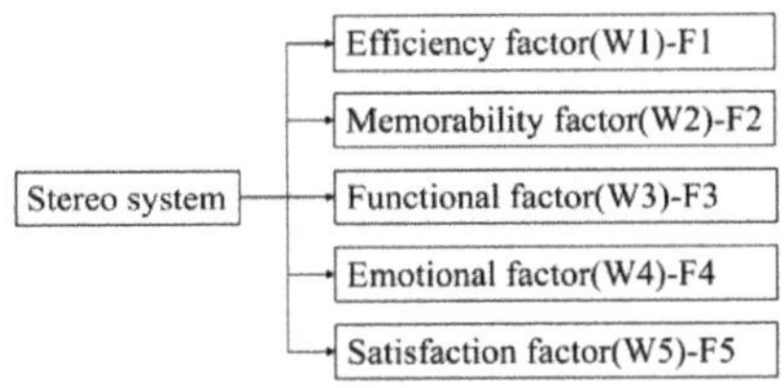

Figure 13. Hierarchy of five items of a stereo system.

Table 6. Weights of five items.

Item	F1	F2	F3	F4	F5	K_i	$\sum K_i$	W_i
F1	\	2	1	2	1	6	40	0.15
F2	2	\	1	2	1	6	40	0.15
F3	3	3	\	3	1	10	40	0.25
F4	2	2	1	\	2	7	40	0.175
F5	3	3	3	2	\	11	40	0.275

All of the weights can be calculated by the above-mentioned method and the results are summarized in Table 7.

Table 7. Weights of all items.

Efficiency Factor (W1) 0.15	Precision of the audio source button (W11)	0.375
	Message reception promptness (W12)	0.25
	Message reception clarity (W13)	0.25
	Fast taking/placing CD (W14)	0.125
Memorability Factor (W2) 0.15	Visibility of button design (W21)	0.375
	Complexity of interface design (W22)	0.125
	CD-ROM's location and external style and dimensions (W23)	0.29
	CD-ROM's way of mechanical operation (W24)	0.21
Functional Factor (W3) 0.25	Other ways of placement (W31)	0.29
	Adequacy of the product's spatial arrangement (W32)	0.29
	Angle of product's speakers and the consistency of sound (W33)	0.17
	Easy moving (W34)	0.25
Emotional Factor (W4) 0.175	Novelty of style (W41)	0.25
	Preference of style design (W42)	0.25
	Audible added value of the style (W43)	0.17
	Enhanced overall atmosphere when listening to music (W44)	0.33
Satisfaction Factor (W5) 0.275	Proportion and arrangement of the product's screen and speakers (W51)	0.33
	The frequency of using other functions in addition to CD (W52)	0.25
	Durability (W53)	0.25
	Cleanness (W54)	0.17

4.3.5. Measuring the Membership Functions

In order to determine the participants' subjective degree of satisfaction on these four design ideas, the membership values of all items need to be measured. The definition of the fuzzy evaluation set is as follows:

$$\underset{\sim}{V} = \left\{ \frac{0}{\text{Strongly disagree}}, \frac{0.25}{\text{Disagree}}, \frac{0.5}{\text{Neither agree nor disagree}}, \frac{0.75}{\text{Agree}}, \frac{1}{\text{Strongly agree}} \right\} \quad (3)$$

After that, the average membership values of each child item on the design ideas can be obtained from the values that were calculated from the questionnaire. The results are shown in Table 8.

Table 8. Average measured values from the 30 questionnaire copies.

	Idea 1	Idea 2	Idea 3	Idea 4
Precision of the audio source button (W11)	0.650	0.517	0.783	0.333
Message reception promptness (W12)	0.692	0.683	0.658	0.625
Message reception clarity (W13)	0.583	0.525	0.708	0.525
Fast taking/placing CD (W14)	0.525	0.358	0.533	0.842
Visibility of button design (W21)	0.683	0.425	0.733	0.500
Complexity of interface design (W22)	0.183	0.450	0.350	0.492
CD-ROM's location and external style and dimensions (W23)	0.333	0.608	0.308	0.267
CD-ROM's way of mechanical operation (W24)	0.517	0.758	0.383	0.383
Other ways of placement (W31)	0.517	0.588	0.677	0.333
Adequacy of the product's spatial arrangement (W32)	0.300	0.642	0.783	0.475
Angle of product's speakers and the consistency of sound (W33)	0.483	0.792	0.567	0.525
Easy moving (W34)	0.242	0.567	0.525	0.467
Novelty of style (W41)	0.300	0.842	0.417	0.508
Preference of style design (W42)	0.533	0.755	0.617	0.422
Audible added value of the style (W43)	0.300	0.383	0.333	0.442
Enhanced overall atmosphere when listening to music (W44)	0.608	0.667	0.683	0.533
Proportion and arrangement of the product's screen and speakers (W51)	0.642	0.583	0.667	0.408
The frequency of using other functions in addition to CD (W52)	0.533	0.658	0.700	0.492
Durability (W53)	0.700	0.400	0.733	0.683
Cleanness (W54)	0.617	0.567	0.658	0.525

After that, the influence of the efficiency factors (accuracy of music source button, promptness of message reception, clarity of message reception, time for taking/placing CD) at the lower level on the efficiency factors at the higher levels. The influence of the four items on the four prototype stereo systems can be determined by the same approach from the lower levels up to the higher levels.

1. Efficiency factors:

$$[E] = [w1\ w2\ w3\ w4] * [E] = [w1\ w2\ w3\ w4] * \begin{bmatrix} r11 & r11 & r11 & r11 \\ r12 & r12 & r12 & r12 \\ r13 & r13 & r13 & r13 \\ r14 & r14 & r14 & r14 \end{bmatrix}$$
$$= [0.375\quad 0.25\quad 0.25\quad 0.125] * \begin{bmatrix} 0.650 & 0.517 & 0.783 & 0.333 \\ 0.692 & 0.683 & 0.658 & 0.625 \\ 0.583 & 0.525 & 0.708 & 0.525 \\ 0.525 & 0.358 & 0.533 & 0.842 \end{bmatrix} \tag{4}$$
$$= [0.628\quad 0.628\quad 0.702\quad 0.518]$$

The membership values of efficiency factors are shown in Figure 14. It is known from the calculation of the membership functions that, Idea 3 (0.702) had the greatest influence from the aspect of the efficiency factor, followed by Idea 1, Idea 2 and Idea 4.

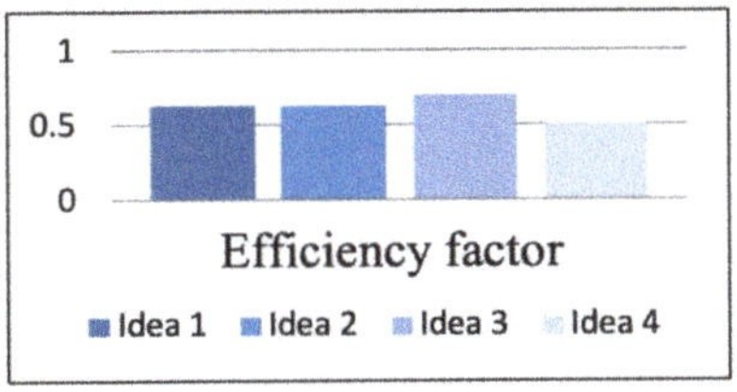

Figure 14. Membership values of efficiency factors.

The membership values of memorability factors, functional factors, emotional factors, and satisfaction factors are respectively shown in Figures 15–18 which indicated that in general Idea 2 had the great influence than others except from the aspect of the satisfaction factor.

2. Memorability factors:

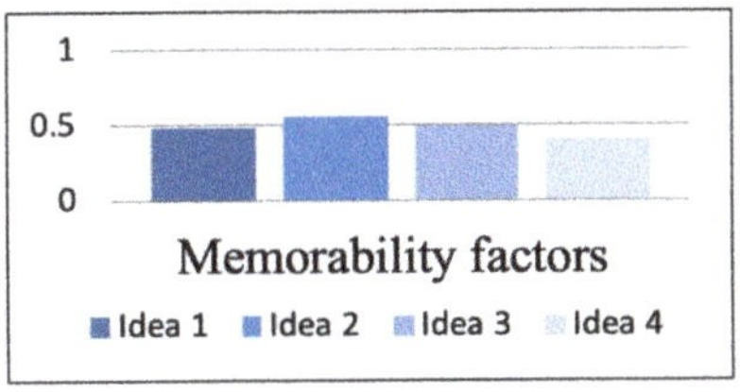

Figure 15. Membership values of memorability factors.

3. Functional factors:

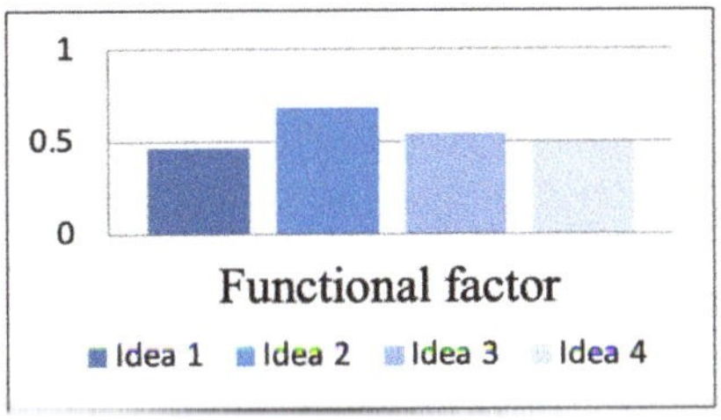

Figure 16. Membership values of functional factors.

4. Emotional factors:

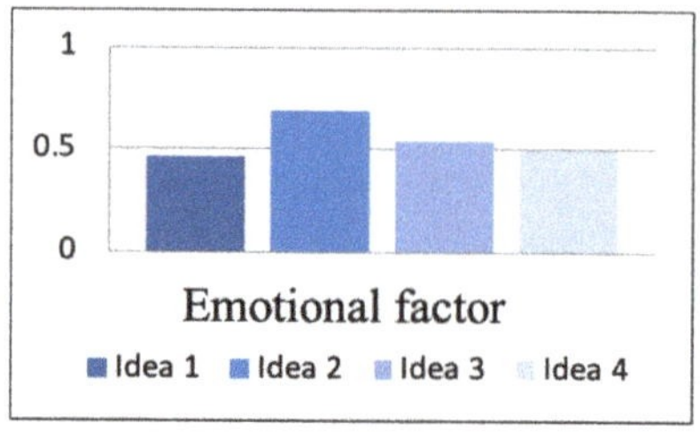

Figure 17. Membership values of emotional factors.

5. Satisfaction factors:

Figure 18. Membership values of satisfaction factors.

4.4. Evaluating the 'Assembleability' of Design Schemes

A product design is considered not good enough if it cannot be manufactured and assembled in an efficient way. The implementation of the design for assembly (DFA) approach during product design is to improve the productivity. The application of DFA presents several advantages as follows:

1. Excessive components in a product can be reduced.
2. The assembly efficiency can be improved.
3. The assembly time of a product can be reduced.
4. Members in a design team can spend less time on communication.
5. The number of components that are used in a product can be reduced after a redesign.

During the first stage, the main task is to determine the minimum number of components. The calculation of the minimum number of components is determined by considering whether the components can deliver the required functions for the product. It is advised to carry out a discussion on the four questions as follows:

6. Is the component required to take relative motion to other components during operation?
7. Is the component separate to other components that are already assembled or does it require different types of materials?
8. When assembling a component or carrying out maintenance, is it required separating a component from other components that are already assembled?
9. Is the component the only one that can deliver the desired function?

If the answer to any of the above questions is yes, the component should exist in a standalone way. During the second stage, the main task is to analyze and measure the manual handling and manual insertion of each component:

10. Manual handling: This deals with the motions including grabbing, transporting and positioning components and their direction.

11. Manual insertion: This involves the motions of placing a component and adding it to the tooling to form an assembly.
12. Aiming at manual assembly, find a two-digit manual code and a two-digit placing motion code.
13. Carrying out the evaluation according to the manual handling and manual insertion tables that were proposed by Boothroyd and Dewhurst [17] as follows.

4.4.1. Structure Tree of Design Schemes

A graphical description is used to build up the assembly sequence of the assembling process. Each design idea is presented in exploded view drawing. The dimensions of each component are marked in the three-view drawing of each component. The sequence is shown in Figures 19–22.

Idea 1:

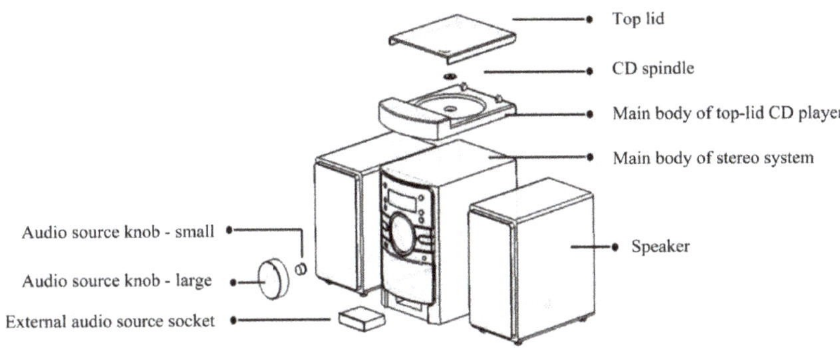

Figure 19. Exploded view drawing of the assembly of Idea 1.

Idea 2:

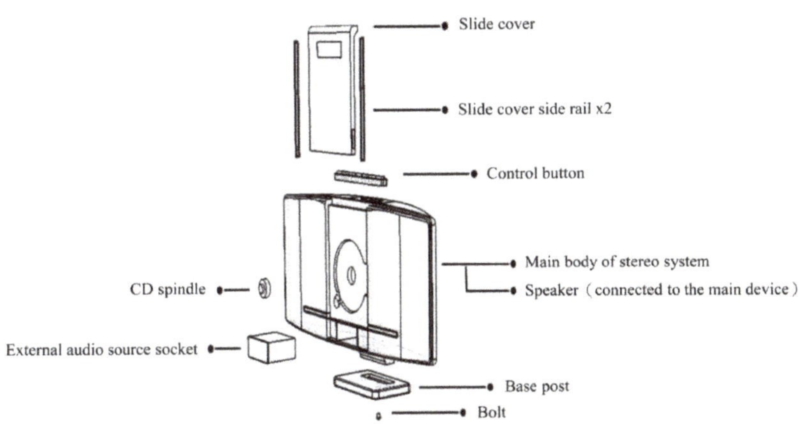

Figure 20. Exploded view drawing of the assembly of Idea 2.

Idea 3:

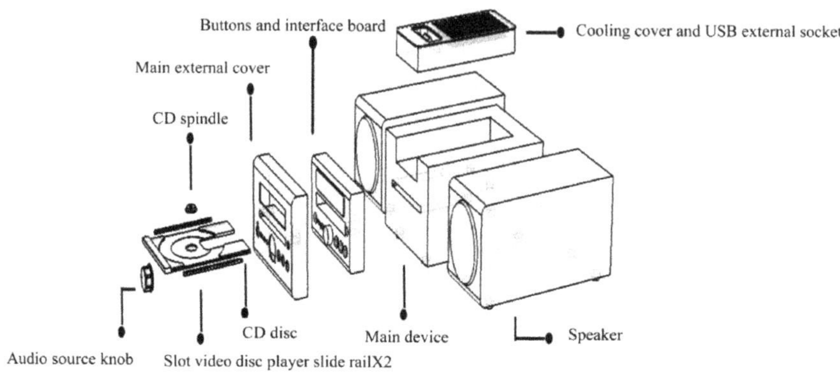

Figure 21. Exploded view drawing of the assembly of Idea 3.

Idea 4:

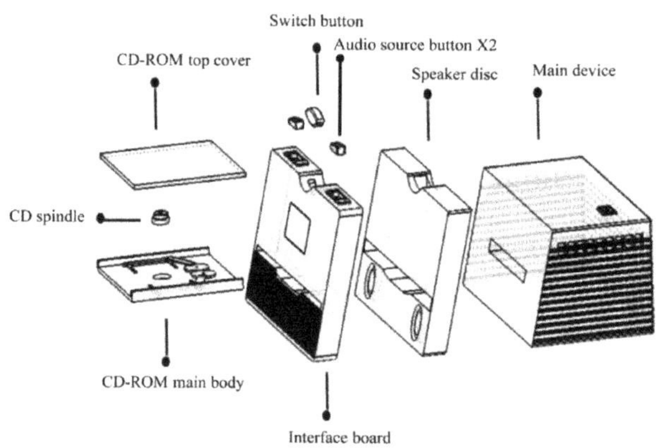

Figure 22. Exploded view drawing of the assembly of Idea 4.

In order to calculate the assembly efficiency, assembly time and assembly cost according to the DFMA table, each component of these four design ideas was displayed separately and its length, width and height are marked in the drawing.

4.4.2. DFMA Analysis of the B and D of Design Ideas

After the dimensions of each component were determined in the exploded view drawing, the data was filled into the DFMA table. At the completion of this table, the total assembly time, total assembly cost and the necessity of each component of each design concept could be determined.

4.5. Analysis of the 'Assemblability' Evaluation Results

After the assembly time and cost of these four design ideas were determined, the change rate of the component numbers and that of the assembly time are shown in Table 9.

Table 9. Comparison of assembly efficiency of each component.

	Idea 1	Idea 2	Idea 3	Idea 4
Percentage of change in the component numbers (%)	Reduced by 12%	Reduced by 33%	Reduced by 22%	Reduced by 9%
Percentage of change in the assembly time (%)	Reduced by 10%	Reduced by 24%	Reduced by 17%	Reduced by 14%

It is known from Table 9 that the improvement of Idea 2 is the optimal one among these four design ideas and it is more suitable for improvement from the assembly efficiency. The optimal sequence of the assembly efficiency of these four design ideas is as follows:

Idea 1 ≥ Idea4 > Idea3 > Idea 2

After that, the next section is to investigate the consideration from the usability aspect. The sequence that was obtained from the engineering aspect is included for a comprehensive investigation.

5. Correlation Between Assembly Design and Usability Operation

The influence of the items at the upper level on the child levels can be determined from the procedure in the above section. In order to make the weights at the upper level suitable for the decision of the design scheme replacement, the application of βvalue should be discussed as follows [18].

5.1. Definition of β Value

A suitable β value is affected by each item's weighting function (*Wi*) and fuzzy probability (*rij*). Therefore, β value often varies with the weighting function (*Wi*) and fuzzy probability (*rij*).

The method of M(*,+,β) is also called the generalized weighted mean method. The definition is as follows:

$$e_j = \{\sum_{i=1}^{m} w_i \times r_{ij}\beta\} \times \frac{1}{\beta},\ \mathrm{j} = 1,2, \ldots, \mathrm{n}. \quad (5)$$

An additional index β is adopted and the β value is in the range of $-\infty$ to $+\infty$. The β value is determined by the designers according to attribute of each problem.

Under the condition that W1 = 0.15, W2 = 0.15, W3 = 0.25, W4 = 0.175, W5 = 0.275 for the weighting functions, the variation in the membership function can be determined.

5.2. Influence of the Weighting Function

In order to test the difference between the weighting functions, the results of these five different weighting functions (W1, W2, W3, W4, W5) were reviewed as follows:

(a) The efficiency factor (W1) is increased from 0.15 to 0.4 and W2 = 0.15, W3 = 0.15, W4 = 0.15, W5 = 0.15.
(b) The memorability factor (W2) is increased from 0.15 to 0.4 and W1 = 0.15, W3 = 0.15, W4 = 0.15, W5 = 0.15.
(c) The functional factor (W3) is increased from 0.25 to 0.4 and W1 = 0.15, W2 = 0.15, W4 = 0.15, W5 = 0.15.
(d) The emotional factor (W4) is increased from 0.175 to 0.4 and W1 = 0.15, W2 = 0.15, W3 = 0.15, W5 = 0.15.
(e) The satisfaction factor (W5) is increased from 0.275 to 0.4 and W1 = 0.15, W2 = 0.15, W3 = 0.15, W4 = 0.15.

5.3. Fuzzy Average Predicted Values

In order to make decisions on a single β value, the detailed fuzzy average predicted values can be determined for decision-making. They can be expressed by the following equation [19].

$$S = \int_{\beta 1}^{\beta 2} ej(\beta)d\beta / \int_{\beta 1}^{\beta 2} d\beta \quad (6)$$

The $\beta1$ and $\beta2$ values are the upper and lower limits and a decision is made between these two values. ej (β) indicated the membership function of the jth design idea. The values are shown in Table 10 as follows.

Table 10. Fuzzy average predicted values of four design ideas.

Weight	Mean Fuzzy Probability			
	Idea 1	**Idea 2**	**Idea 3**	**Idea 4**
W1 = 0.4 W2 = 0.15 W3 = 0.15 W4 = 0.15 sW5 = 0.15	0.612	0.66 *	0.63	0.48
W1 = 0.15 W2 = 0.4 W3 =0.15 W4 = 0.15 W5 = 0.15	0.594	0.646 *	0.573	0.589
W1 = 0.15 W2 = 0.15 W3 = 0.4 W4 = 0.15 W5 = 0.15	0.592	0.687 *	0.555	0.581
W1 = 0.15 W2 = 0.15 W3 = 0.15 W4 = 0.4 W5 = 0.15	0.632	0.699 *	0.596	0.531
W1 = 0.15 W2 = 0.15 W3 = 0.15 W4 = 0.15 W5 = 0.4	0.631	0.682 *	0.586	0.491

(* indicated the highest value of correlation).

It is known from the above table that Idea 2 has the highest membership value. Therefore, by strengthening the weighting relationship between five different considerations, it can be observed that Idea 2 is the optimal and the most suitable decision. The overall product is shown in Figure 23 as follows.

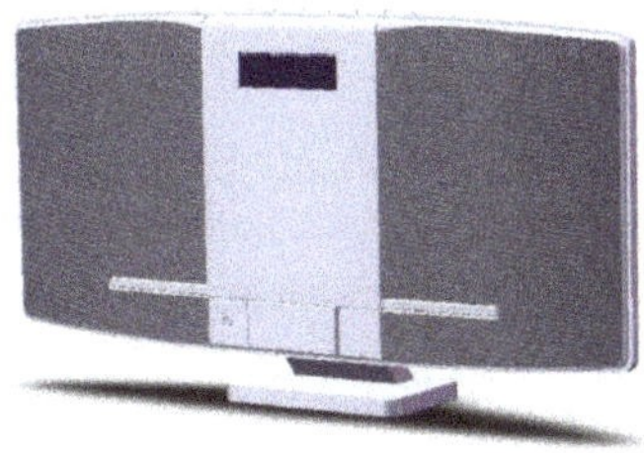

Figure 23. Idea 2 as the optimal idea.

5.4. Results and Analysis of the Correlation Between Assembly and Usage

Under the condition of weighting functions of W1 = 0.15, W2 = 0.15, W3 = 0.25, W4 = 0.175, W5 = 0.275, the variation in the membership functions can be determined. It is known from the consideration of usability that, the ranking of the design ideas according to the participants' preference is as follows.

Idea 2 > Idea 1 > Idea3 > Idea4

Moreover, the comparison to the sequence of assembly design is shown in Table 11.

Table 11. Comparison of the correlation between assembly design and usability.

	Ranking			
Complexity of usability operation	Idea 2	Idea 1	Idea3	Idea4
Assembly design	Idea 1	Idea4	Idea3	Idea 2

It is known from Table 11 that Idea 1 and Idea 3 are ranked higher from the aspect of engineering and usability. Therefore, if a company is planning to roll out new stereo systems, it is advised to extract some design elements that meet consumer demands from Idea 1 and Idea 3 for the consideration of product innovation.

6. Conclusions and Recommendation

An evaluation system for assembly design and usability operation complexity was proposed in this study. The problems that are related to product designs were handled by computers. The decision-making on the assembly sequence was assisted by the calculation of matrices according to the principles of assembly design. This approach can reduce the errors and problems due to personal factors of the operators. The potential interference problems between components of a product can also be eliminated by computer-aided analysis during the assembly process.

The fuzzy theory was applied to the decision-making from the concept development stage to sketches and finally to the product mock-ups. The participants were allowed to simulate the mechanical movements of these four different design ideas. From the scores that were given by the participants, the optimal design concept can be determined. The design ideas were ranked from the consideration of usability without considering much of the engineering aspect. Therefore, the engineering evaluation was accomplished by DFA, which can calculate the assembly efficiency, assembly time and cost of these four design ideas. From the consideration of the DFA engineering aspect, the ranking of these four design ideas can also be determined. This approach can not only improve the usability but can also enhance a product's value from the engineering aspect for the reference of follow-up research.

The development process of each product involves the joint endeavor of many departments. Therefore, when constructing the decision model of a product, it is required to correlate the design logics prior to and after each process. The considerations include the development basis of sketches, the parametric configuration of modeling, the simulation of mechanisms of the assembly and the questionnaire design. The correlation between weighting functions and the transformation of the questionnaire into engineering parameters for further application are also important and should follow the strict rule.

In order to respond to the rapid changes in market demands, a product designer needs to develop a design and production approach that can adapt to the rapid changes in product styles. Therefore, a product designer is required to create designs that present different ways of using or different appearances in order to satisfy various consumer demands such as various types of coffee machines and speaker systems. Meanwhile, in order to broaden the usage range for consumers, a product design approach is expected to allow designers to increase or decrease the number of functional components while the product appearance remains the same. Alternatively, a different product style can be created by altering the pattern of one of the internal components such as various types of desktop computers. Under the market mechanism, a product designer needs to carry out the development of product variability in order to enhance a product's competitiveness on the market, by satisfying different consumer usage models or different preferences.

This study aims to connect the manufacturing end to the user end so that a simple and basic assembly planning can be realized during the earlier design stage. Via the concurrent engineering concept, the product design quality and efficiency can be enhanced. Moreover, via the engineering-oriented assembly consideration and the user-oriented consideration, a designer is allowed to consider design feasibility during the earlier design stage so that the final product can be accepted and recognized by general users.

Funding: This work was supported by the Ministry of Science and Technology of the Republic of China under grant MOST 107-2635-E-468-001.

Conflicts of Interest: The authors declare no conflict of interest.

References

1. Lo, C.H.; Ko, Y.C.; Hsiao, S.W. A methodology for high-performance operating interface design. *Concurr. Eng. Res. Appl.* **2015**, *23*, 110–123. [CrossRef]
2. Ko, Y.C. A Study of the Hand Operating Thresholds during the Usage of an Electric Steam Iron by College Students. *Eurasia J. Math. Sci. Technol. Educ.* **2017**, *13*, 6325–6335. [CrossRef]
3. Ko, Y.C.; Lo, C.H.; Hsiao, S.W. Interface Design Optimization by an Improved Operating Model for College Students. *Eurasia J. Math. Sci. Technol. Educ.* **2017**, *13*, 2601–2625. [CrossRef]
4. Kwahk, J.; Han, S.H. A methodology for evaluating the usability of audiovisual consumer electronic products. *Appl. Ergon.* **2002**, *33*, 419–431. [CrossRef]
5. Lee, S.; Heo, G.; Chang, S.H. Prediction of the human response time with the similarity and quantity of information. *Reliab. Eng. Syst. Saf.* **2006**, *91*, 728–734. [CrossRef]
6. Lee, Y.S.; Hong, S.W.; Smith-Jackson, T.L.; Nussbaum, M.A.; Tomioka, K. Systematic evaluation methodology for cell phone user interfaces. *Interact. Comput.* **2006**, *18*, 304–325. [CrossRef]
7. Heo, J.; Ham, D.H.; Park, S.; Song, C.; Yoon, W.C. A framework for evaluating the usability of mobile phones based on multi-level, hierarchical model of usability factors. *Interact. Comput.* **2009**, *21*, 263–275. [CrossRef]
8. Jin, B.S.; Ji, Y.G. Usability risk level evaluation for physical user interface of mobile phone. *Comput. Ind.* **2010**, *61*, 350–363. [CrossRef]
9. Chou, J.R. An Empirical Study of User Experience on Touch Mice. *Eurasia J. Math. Sci. Technol. Educ.* **2016**, *12*, 2875–2885. [CrossRef]
10. Hsiao, S.W. Fuzzy set theory on car-color design Color. *Res. Appl.* **1994**, *19*, 202–213. [CrossRef]
11. Barbosa, G.F.; Carvalho, J. Design for Manufacturing and Assembly methodology applied to aircrafts design and manufacturing. *IFAC Proc. Vol.* **2013**, *46*, 116–121.
12. Moultrie, J.; Maier, A.M. A simplified approach to design for assembly. *J. Eng. Des.* **2014**, *25*, 44–63. [CrossRef]
13. Tiwari, M.K.; Battaïa, O.; Dolgui, A.; Heragu, S.; Meerkov, S. Design for manufacturing and assembly/disassembly: Joint design of products and production systems. *Int. J. Prod. Res.* **2017**, *56*, 7181–7189.

14. Sossou, G.; Demoly, F.; Montavon, G.; Gomes, S. An additive manufacturing oriented design approach to mechanical assemblies. *J. Comput. Des. Eng.* **2018**, *5*, 3–18. [CrossRef]
15. Boothroyd, G.; Dewhurst, P. *Design for Assembly: A Designer's Handbook*; Departement of Mechanical Engineering, University of Massachusetts: Wakerfield, RI, USA, 1983.
16. Norman, D.A. *The Design of Every Things*; MIT Press, Massachusetts Institute of Technology: Cambridge, MA, USA, 1988.
17. Boothroyd, G.; Dewhurst, P.; Knight, W. Product Design for Manufacture and Assembly. *Comput. Aided Des.* **1994**, *26*, 505–520. [CrossRef]
18. Tuttle, B.L. Design for function: A Cornerstone for DFMA. In Proceedings of the International Forum on Product Design for Manufacture and Assembly, Newport, RI, USA, 10–11 June 1991.
19. Cross, N. *Engineering Design Methods*; Wiley: New York, NY, USA, 1994.
20. Saaty, T.L. *How to Make a Decision: The Analytic Hierarchy Process*; Joseph, M., Ed.; McGraw-Hill: New York, NY, USA, 1980.
21. Chien, C.C.; Feng, M.; Chu, C.M. *Methodology in Industrial Design*; Beijing University Press: Beijing, China, 1993.

applied sciences

MDPI

Article

An LSTM-Based Deep Learning Approach for Classifying Malicious Traffic at the Packet Level

Ren-Hung Hwang *, Min-Chun Peng, Van-Linh Nguyen and Yu-Lun Chang

Department of Computer Science and Information Engineering, National Chung Cheng University, Chiayi 62102, Taiwan
* Correspondence: rhhwang@cs.ccu.edu.tw

Received: 12 July 2019; Accepted: 14 August 2019; Published: 19 August 2019

Abstract: Recently, deep learning has been successfully applied to network security assessments and intrusion detection systems (IDSs) with various breakthroughs such as using Convolutional Neural Networks (CNN) and Long Short-Term Memory (LSTM) to classify malicious traffic. However, these state-of-the-art systems also face tremendous challenges to satisfy real-time analysis requirements due to the major delay of the flow-based data preprocessing, i.e., requiring time for accumulating the packets into particular flows and then extracting features. If detecting malicious traffic can be done at the packet level, detecting time will be significantly reduced, which makes the online real-time malicious traffic detection based on deep learning technologies become very promising. With the goal of accelerating the whole detection process by considering a packet level classification, which has not been studied in the literature, in this research, we propose a novel approach in building the malicious classification system with the primary support of word embedding and the LSTM model. Specifically, we propose a novel word embedding mechanism to extract packet semantic meanings and adopt LSTM to learn the temporal relation among fields in the packet header and for further classifying whether an incoming packet is normal or a part of malicious traffic. The evaluation results on ISCX2012, USTC-TFC2016, IoT dataset from Robert Gordon University and IoT dataset collected on our Mirai Botnet show that our approach is competitive to the prior literature which detects malicious traffic at the flow level. While the network traffic is booming year by year, our first attempt can inspire the research community to exploit the advantages of deep learning to build effective IDSs without suffering significant detection delay.

Keywords: deep learning for network security; long short-term memory; malicious traffic classification

1. Introduction

In a gigantic connected world like the Internet-of-Things (IoT), protecting the network from network attacks may require a comprehensive approach to defeat while under the limitation of the existing resources, e.g., the capacity of the gateway or edge server. The task is getting much harder with the rapid increase of the volume of attacks such as the distributed denial of service (DDoS) from IoT devices. For example, a Mirai-based botnet launches one of the biggest DDoS attacks in history directed at Dyn in October of 2016. The attack surpassed 1 Tbps at peak and created disruption for many major sites, including AirBnB, Netflix, PayPal, Visa, Amazon, the New York Times and so on. Currently, the variants of Mirai malware such as Satori and Miori are still storming to the network with the new records of traffic volume towards the victim, including the systems of the enterprise companies. In practice, the attacker often handles a large number of the botnets of injected IoT devices to launch such an attack. As a result, detecting the malware traffic in the early period of distributing the malicious code can significantly help to prevent the malware from becoming widespread, and mitigate the attack magnitude.

Over the decades, the malicious traffic detection system is the top topic and considered the most in the intrusion detection system (IDS) field. Thus far, the most popular approach of such IDSs is to build a classification mechanism to distinguish the incoming traffic into benign and malicious classes. An IDS can use the signature-based or rule-based approach to identify abnormal traffic. Apparently, the major drawbacks of these traditional IDSs include: (1) requiring many efforts, e.g., labor, to maintain the signature database and updates; (2) becoming ineffective for maintaining such database if the new traffic types or attack variances increasingly spike in a short time. In addition, many adversaries still have certain advantages because exploiting vulnerabilities in the systems of millions of codes, including the zero-day ones available on the darknet, may not require much effort. When the ecosystem of Internet-connected systems expands and the number of IoT devices inside increased rapidly, the attack surface also increases. Consequently, that increase can cause a greater risk of attack and easily overwhelm the update requests. Note that the traffic promises to explode quickly when massive quantities of IoT applications are predicted to flood the market in the next few years, particularly with the readiness of 5G networks. In this case, deep learning promises to be a game changer to ignite new detection approaches. The most benefits of the deep learning approaches are to build a thorough pattern that can highly represent the characteristics of a specific object through auto-learning a large volume of data and species.

Although applying deep learning (DL) in network security has just emerged recently, the topic has received a lot of attention from the research community [1–7] due to the robust auto-learning ability of DL. In addition, the evolution and the increased availability of GPU-processors significantly help to accelerate matrix computations and massive mathematical calculations, thus directly supporting the feasibility of the DL-based approaches. Deep learning for malicious traffic detection is generally categorized into many classes that are primarily based on the built-in network model, e.g., Convolutional Neural Network (CNN), Recurrent Neural Network (RNN) or unsupervised learning such as an auto-encoder. However, existing literature still suffers several critical drawbacks. First, such systems are built on the flow-based approach (i.e., collecting packets of the same flow for a certain period of time and then classifying packets into flows) and evaluated offline. Since the system needs to allocate significant time for accumulating the traffic flow, it is not applicable to an online real-time system. This packet collection time can be extremely high if the flow is likely the part of a long session data. Furthermore, extracting the features for the flow-based detection system also requires a certain period of time. Apparently, this whole traffic profiling process naturally increases the overall time of the detection. Secondly, the flow-based classification can require a lot of resources, e.g., memory and storage, to store and process the accumulated traffic, including the deep checking on the assembled data. Some recent works, as well as our previous work [8], have shown that it is not necessary to examine all packets in a flow to determine whether it is a malicious flow; instead, a few first packets of the flow are enough for the detection purposes.

Based the above observation, in this research, we propose a novel approach to build a packet-based malicious traffic classification framework. Instead of wasting time on waiting for a complete flow of packets and then launching the classification, we use word embedding to learn the similarity in semantics and syntax of header fields a packet and leverage LSTM to learn temporal features of the fields. The goal of the method is to identify the attack only by directly monitoring a packet.

Due to considering the traffic at the packet level, our system has significant advantages to speed up the detection process, e.g., ignore checking a large number of packets in a session. The experimental results show that our method is competitive and prominent better in the term of accuracy, precision, recall rate, and F1-measure as compared to the prior work.

In summary, the main contributions of this study are as follows:

(1) We propose an LSTM-based deep learning approach for the packet-level classification in IDSs. Specifically, our approach can distinguish semantic meanings of malicious traffic by packet, and are thus highly competitive with the flow-based DL approaches while reducing significant time on flow processing.

(2) We propose a novel word-embedding and LSTM model to classify the incoming packet into the normal and abnormal state. The evaluation results of the proposed model on ISCX2012, USTC-TFC2016, IoT data set from Robert Gordon University and IoT data set collected on our Mirai bonet show that the proposed framework can achieve near 100% accuracy and precision in detecting malicious packets.

(3) Our discussion on packet-based deep learning classification and detection promises to provide valuable information and inspire the research community to overcome the remaining challenges, particularly for the target of speeding up the detection process in DL-based IDSs.

The remainder of this paper is organized as follows. Section 2 lists several relevant fundamental concepts and state-of-the-art work of traffic classification and deep learning detection approaches. Section 3 presents our methodology and the detail architecture of the packet classification module in our classification system, along with the data set for evaluation. We show the experimental design and results in Section 4. The conclusions and future work of this paper are summarized in Section 5.

2. Related Work

Deep learning for malicious traffic detection has gained several notable achievements with various network models. For example, the authors in [2] proposed a novel network-based anomaly detection method which extracts behavior snapshots of the network and uses deep autoencoders to detect anomalous network traffic emanating from compromised IoT devices. The method is evaluated on commercial IoT devices infected by authentic botnets such as Mirai and BASHLITE. However, the performance of the work primarily relies on several self-generated synthetic data sets, which may lack the diversity of data exchange. In another research work [3], the authors proposed a malware traffic classification method using CNN by considering traffic data as images. The work is one of the first attempts to apply a representation learning approach for malware traffic classification from raw traffic. Unfortunately, the lack of detecting the unknown attacks and high detection time limits its prospect to deploy in practice. In addition, Li et al. [7] and Yin et al. [6] proposed using an RNN model or a combination with Restricted Boltzmann Machines (RBM) for extracting micro-flow features based on a small set of packets and then directly training with the raw packets. However, they mostly target the detection accuracy and ignore the evaluation of the detection time. Thus far, there are several other state-of-the-art relevant studies such as [9,10]. Generally, the key points of such work are that they all rely on the flow or session-based approach, i.e., indexing the traffic by the flow. Thus far, we have found no attempt to detect the attack traffic at the packet level. Our statistics on the relevant research and our research position are summarized in Table 1.

Table 1. Summary of several state-of-the-art relevant research works and our research location. (Notations: GRU: Gated Recurrent Unit; TSDNN: Shaped Deep Neural Network; QDBP: Quantity Dependent Backpropagation; WEDL: Word Embedding-based Deep Learning; NIDS: Network Intrusion Detection System).

Research	Year	Method	Feature	Shortcomings
M. S. Kim [11]	2004	Formulization	flow-based	For offline detection
R. Fu [12]	2016	LSTM, GRU	flow-based	For offline detection
B J. Radford [9]	2018	LSTM	flow-based	For offline detection
C. Li [7]	2018	RNN, RBM	process packet, flow-based	For offline detection
Y. Chen [4]	2018	TSDNN, QDBP	flow-based, N class	For offline classification
X. Yuan [5]	2017	LSTM, GRU	process packet, flow-based	For offline detection
J. Cui [10]	2018	WEDL-NIDS	process packet, flow-based	For offline detection
Ours	2019	LSTM, Word-embedding	raw data, packet-based	Target for online detection

Recently, word2vector [13,14] and LSTM-based learning models [10] provide a very powerful tool to give packet semantic meanings. This means we now can find a way to provide a thorough pattern for the malicious traffic by deep learning of a large number of raw packets. This particularly helps to reinforce our approach to consider the malicious traffic classification at the packet level. Compared to the prior studies, there are two major differences in our work: (1) the system can identify the malicious traffic by classifying the individual packet, instead of checking the whole packets of a traffic flow; (2) our system can work directly with raw packets, i.e., reading and making detection decisions. To the best of our knowledge, our study is the first work on packet-based malicious traffic detection. In addition, despite the high detection accuracy achieved by prior work for the same attack types, a novel research direction on packet-based classification (i.e., ours) sheds a light toward research on online real-time deep learning based IDSs.

3. Methodology

3.1. Dataset

In deep learning, a lack of a variety of shareable traces dataset can be seen as one of the most obvious obstacles to obtain realistic progress on traffic classification. Many researchers may like creating synthetic datasets through their testbed. The shortcomings of this approach are that their generated traffic may not represent well the real traffic on the Internet, thus the performance evaluation results of the proposed solutions can face the problems of the credibility. Over the decades, the researchers have been recommended to use public famous traffic datasets such as KDD CUP1999 and NSL-KDD to test. Each dataset explicitly provides useful statistics on labeled features and the number of benign and malicious flows as well. However, these datasets do not provide information at the raw traffic level, which is required in our approach. In addition, while the credible dataset from Microsoft Malware Classification Challenge [15] can provide a metadata manifest and hexadecimal representation of the malware's binary content, the missing of packet information in the data, unfortunately, makes it hard to be used in this research.

For the ones matching the requirements, USTC-TFC2016 [3] is one of the most prominent datasets. Table 2 summarizes the statistics of the benign and malware traffic in the dataset. As their statement, there are a total of ten types of malware traffic from public websites which were collected from a real network environment from 2011 to 2015. Along with such malicious traffic, the benign part contains ten types of normal traffic which were collected using IXIA BPS, a professional network traffic simulation equipment. The size of USTC-TFC2016 dataset is 3.71 GB, and the format is pcap.

Table 2. Summary of benign and malware traffic in USTC-TFC2016 dataset. (Notations: SMB: Server Message Block; IM: Instant Message; P2P: Peer-to-Peer).

Benign			Malware	
App Type	**Size (MB)**	**Class**	**Malware Type**	**Size (MB)**
Facetime	2.4	Voice/Video	Tinba	2.55
Skype	4.22	Chat/IM	Zeus	13.4
Bittorent	7.33	P2P	Shifu	57.9
Gmail	9.05	Email/Webmail	Neris	90.1
Outlook	11.1	Email/Webmail	Cridex	94.7
WorldOfWarcraft	14.9	Game	Nsisay	281
MySQL	22.3	Database	Geodo	28.8
FTP	60.2	Data transfer	Miuref	16.3
SMB	1206	Data transfer	Virut	109
Weibo	1618	Social Network	Htbot	83.6

For the Mirai-based DDoS traffic, we use the dataset from Robert Gordon University [16], denoted by Mirai-RGU. This data set contains Mirai botnet traffic such as Scan, Infect, Control, Attack traffic and normal IoT IP Camera traffic. It contains ten types of malicious traffic, include HTTP flood, UDP flood, DNS flood, Mirai traffic, VSE flood, GREIP flood, GREETH flood, TCP ACK flood, TCP SYN flood, and UDPPLAIN flood. The dataset includes features such as Time, Source, Destination, Protocol,

Length, and overall payload information. In addition, we also collected a dataset from the Mirai botnet we have built in the campus of National Chung Cheng University(CCU) (denoted by Mirai-CCU), as shown in Figure 1, to generate four types of attack traffic with much bigger attack magnitude: TCP SYN (41 GB), TCP ACK (2.4 GB), HTTP POST (103 GB), UDP (127.06 GB) in the total of 667 GB attack data.

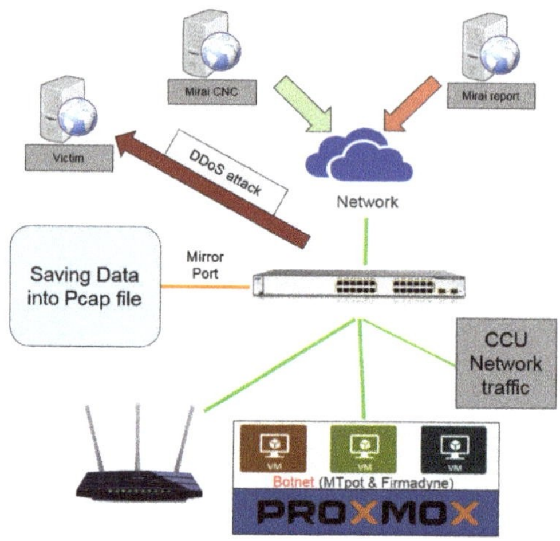

Figure 1. The testbed of our Mirai-based DDoS dataset in the campus of National Chung Cheng University (CCU).

To enrich the dataset for our experiments, we select ISCX2012 [17], which also contains both malicious and benign traffic. ISCX2012 consists of packets collected for seven days. Packets collected in the first and sixth day are normal traffic. In the second and third day, both normal packets and attack packets are collected. In the fourth, fifth, and seventh days, besides the normal traffic, HTTP DoS, DDoS and IRC Botnet, and Brute Force SSH packets are collected, respectively.

3.2. Word Embedding and Data Preprocessing

Our goal is to classify incoming packets into benign or malicious classes without pre-processing the packets into a specific flow. To achieve this target, instead of considering the whole flow (e.g., as a document), we consider each packet (e.g., as a paragraph) and construct the key sentence from each packet, in which each word is a field in the packet header. After that, we apply word embedding [18] to extract semantics and syntax features from this sentence. We choose to consider meanings on the sentence rather than the whole paragraph since the meaning of a paragraph usually can be captured by the key sentence. Here, the order of the fields in each packet (fixed for each packet type) plays the role of resembling some grammar rules which are decisive in building sentence patterns for malicious traffic (signature-based detection) or benign traffic (anomaly detection). Notably, this packet-to-sentence-based model can significantly accelerate the traffic classification, since the behavior and characteristics of one or several first packets can entirely reveal whether their flow is a malicious one.

In general, a field in each packet could be *a byte of the packet header*, *a field of the packet header*, or *a block of the packet payload*. As the initial trial, we consider a field in the packet header as a word and trim a packet to a fixed length of $n = 54$ bytes. Depending on the field length in the packet, the word size can vary. The strict order of the fields in the packet structure constructs a potential grammar rule for the built sentence. Note that the extracting field stage can be done along with the packet reading/decoding (i.e., data pre-processing), thus the resource consumption is quite efficient. In addition, if the length of the packet is less than n bytes, it will be padded with zeros (as shown in

Figure 2). The rationale for examining 54 bytes is that most TCP packets have a 14-byte MAC header, a 20-byte IP header, and a 20-byte TCP header. Table 3 lists the header fields and their length of TCP and UDP packets.

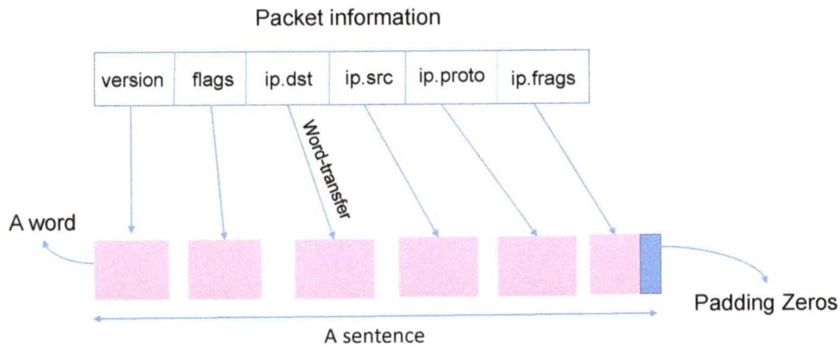

Figure 2. The illustration of the packet-word-transfer mechanism.

Table 3. Packet header(tcp:25, udp: 19) fields, extended fields (tcp: 33, udp: 33) and the length of the fields.

Header	Extended Fields	Details
Ether header (3)	extended to 7 fields	ether.dst(2 × 3) , ether.src(2 × 3), ether.type
IP header (12)	extended to 14 fields	ip.version, ip.ihl, ip.tos, ip.len, ip.id, ip.flags, ip.frag, ip.ttl, ip.proto, ip.chksum, ip.src(2 × 2), ip.dst(2 × 2)
TCP header (10)	extended to 12 fields	tcp.sport, tcp.dport, tcp.seq(2 × 2), tcp.ack(2 × 2), tcp.dataofs, tcp.reserved, tcp.flags, tcp.window, tcp.chksum, tcp.ugptr
UDP header (4)	extended to 12 fields	udp.sport, udp.dpot, udp.len, udp.chksum, 0, 0, 0, 0, 0, 0, 0, 0

After applying the word-embedding technique to fields in the packet header, each header field is embedded, (i.e., to integer number format based on their index in a dictionary of all words) and reshaped, (i.e., dimension) and put to the LSTM-based training model for performing the classification task. In order to understand each attack type and remain consistent with the order of the fields in the packet header, it is important to maintain the sequence order of fields and a consistent word dictionary. Finally, the word size (hyperparameter) selection and further the word-embedding strategy are adjustable and can be specified by the deployment environment and IoT applications under the system's protection umbrella.

3.3. Classification

The classification mechanism is the next important part of the system. As illustrated in Figures 3–5, our packet-based traffic classification framework consists of three modules: packet pre-processing module, word embedding module, and LSTM module. However, to perform training in the LSTM module, we first need to label the benign/malicious mark for the packets in the dataset, e.g., ISCX2012 [17], USTC-TFC2016 [3] and Mirai-RGU [16]. Based on the label of the flows (well-labeled in ISCX2012 and USTC-TFC2016 dataset) and the packets (well-labeled in Mirai-RGU dataset), we mark the label of the corresponding packet (malicious/benign) according to the flow's label that the packet belongs to.

For the LSTM module, we build standard LSTM cells, in which each cell consists of input gate, output gate, and forget gate. The sigmoid function is used as the activation function (α). Our proposed framework, as shown in Figure 4, consists of three LSTM layers with dropout. Total parameters of the first layer (dimension = 128, without dropout) exceed 4 million (4,194,304).

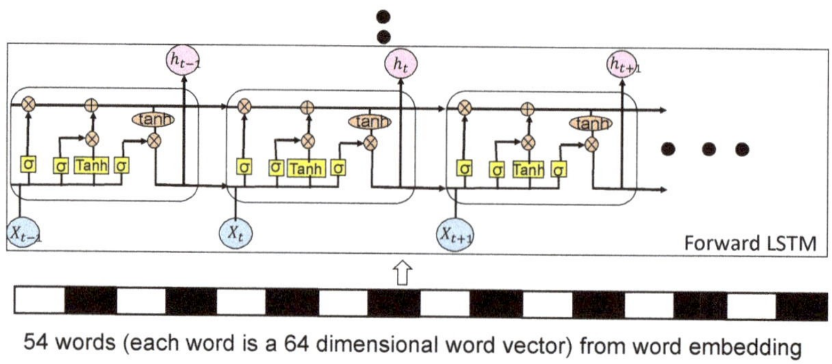

Figure 3. The illustration of the packet-word-transfer and classification module in the proposal. Packet-work-transfer module is at the input data layer.

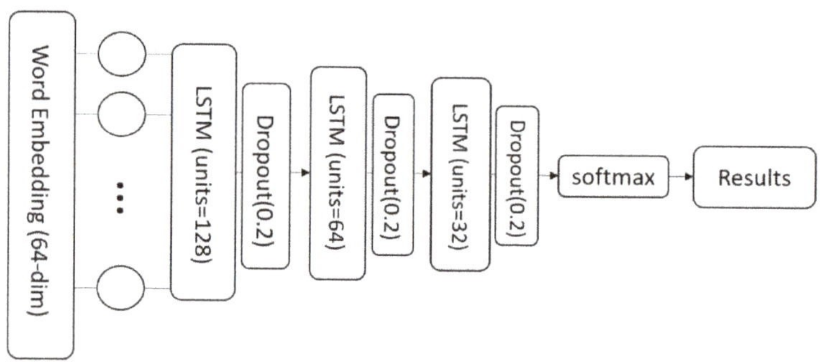

Figure 4. The illustration of the full network model.

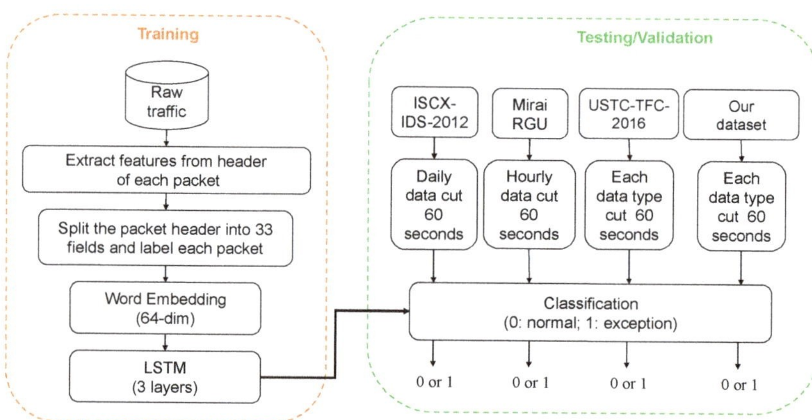

Figure 5. The illustration of the workflow of the training and testing/validation stage. Data pre-processing and packet labelling for training are done at the training stage. Training/Testing is done at the pre-processing data and with the ratio is 9:1. Validation is performed on the random samples (60 s/pcap).

The pseudo-code of our algorithm and the processing flow is illustrated in Algorithm 1. The preprocessing phase adjusts features to ensure data representation is suitable for the used algorithms, i.e., parsing the packet and converting it to the translated word. The new dataset of translated Words is in the format of the dataset of integer number. The translated dataset is then split into two parts: Training Data and Testing Data, respectively. Training stage starts on the Training Data and running the LSTM three-layer model. Note that the dimension from input data are required to reshape, e.g., 64, before performing the training and testing. The dropout rate is flexibly set and can be tuned up to 0.5. The loss function is based on binary cross entropy and RMSProp optimizer is selected for the learning rate adjustment method. A dense output layer with softmax is added to the model. The model is then compiled with the binary cross-entropy as the loss function and the Adam optimiser over total of 200 iterations. Finally, the model is testing on the Testing Data to determine the effectiveness of the model in terms of accuracy, precision, recall, f1-score, and loss (defined later in the next section).

Algorithm 1: Algorithm for packet-based traffic classification

Data: Sequence of raw packets from network

Result: Accuracy, precision, recall, f1-score, FAR, loss

```
dictionary = array() % Create the mapping word to the index array
translatedWords = null; %variable to store the values of words to converted to integer format;
while true do
    Parse the packet;
    Extract 54-byte from each packet
    if packet length < 54 then
        Pad zeros;
    end if
    Parse each byte of the extracted data as a word; words defines the words extracted from the
    packet;
    for {i = 1; i < count(words); i++} do
        if Word in dictionary, dictionary(words[i]) then
            index = dictionary(words[i]) % Pulling index of the words
        else
            dictionary[] = words[i]; %Update the dictionary with the new word;
            index = dictionary(words[i])
        end if
        concat(translatedWords, index)
    end for
end while% return dataset
Split Training and Test based on 90/10 (TrainingData, TestingData)
Train and Validate (TrainingData, TestingData)
Reshaping translatedWords to an output of 64-dimensional word vector;
Input the 64-dimensional word vector to the first layer LSTM;
Dropout;
Feedforward to the second layer LSTM;
Dropout;
Feedforward to the third layer LSTM;
Dropout;
Prepare input for mini-batch (100 packets);
Apply RMSProp optimizer;
Use softmax to output 0 or 1 to the model;
Use binary cross entropy as loss function
for epoch = 1; epoch < 200; epoch++ do
    Evaluate Loss, Validation Loss
    Evaluate Accuracy and Validation Accuracy
end for
```

Training and validation often cost significant time regarding the system evaluation. In fact, the dataset size after converting (transferring to integer format) can be smaller in storage than that of the original text/string as in the packets, since the integer format may cost few spaces than the text data type. However, the dimension unlikely reduces. As a result, selecting word size may dramatically affect the training time. Smaller word size reduces the size of dictionary, but also causes more training data and running time. In this research, we set the maximum word size to 2 bytes based on the consideration of the length of most fields in the packet while limiting the number of words in the dictionary to be less than 65,536.

4. Evaluation Results

Similar to most existing deep learning research, our proposed classification model has been implemented using TensorFlow/Keras (version 2.2.4, Google, Mountain View, CA, USA). All of our evaluations have performed on the GPU-enabled TensorFlow which is running on a 64-bit Ubuntu 18.04.2 LTS server with an Intel(R) Xeon(R) E5-2650 2.2 GHz processor, 32 GB RAM, and an NVIDIA RTX Tesla K80.

To perform our evaluations, we have used the ISCX2012, USTC-TFC2016, Mirai-RGU and Mirai-CCU datasets. During the training and testing stages, we try to include hundreds to thousands of thousands of packets while balancing the benign and malicious traffic. At the validation stage, we run the trained model on the packets which is extracted randomly per 60 s from the selected datasets, i.e., packets in the consecutive 60 s in the original dataset are extracted while maintaining their temporal order. At this stage, the original (real) traffic is presented to the trained model without any manipulation or balancing of two types of traffic. Since our approach aims to gain a significant reduction in processing time, we have many interests in classifying the incoming packet whether it is malicious or not, instead of considering the attack type in detail. In practice, if a malicious packet is detected by the proposed system, it can raise an alarm to the network administrator and direct the packet for some offline computational intensive traffic classification systems. Therefore, in the following section, we define several measurements to evaluate the performance of the proposed solution in the term of a binary classifier. All of these metrics are all derived from the four values found in the confusion matrix in Table 4. The traffic proportion of training/testing/validation configuration of four datasets is listed in Tables 5–8.

Table 4. Confusion matrix.

		Predicted Classs	
		Malicious	**Benign**
Ground truth	Malicious	True Positive (TP)	False Negative (FN)
	Benign	False Position (FP)	True Negative (TN)

where:

- True Positive (TP)—Attack packet that is correctly classified as an attack.
- False Positive (FP)—Benign packet that is incorrectly classified as an attack.
- True Negative (TN)—Benign packet that is correctly classified as normal.
- False Negative (FN)—Attack packet that is incorrectly classified as normal.

The accuracy (Equation (1)) measures the proportion of the total number of correct classifications:

$$Accuracy = \frac{TP + TN}{TP + TN + FP + FN}. \tag{1}$$

The precision implies the number of correct classifications that is influenced by the incorrect classification (Equation (2)):

$$Precision = \frac{TP}{TP + FP}. \tag{2}$$

The recall (Equation (3)) measures the number of correct classifications that is considered with the number of missed entries:

$$Recall = \frac{TP}{TP + FN}. \quad (3)$$

The false alarm or fair (Equation (4)) measures the proportion of benign packets incorrectly classified as malicious:

$$Falsealarmrate(FAR) = \frac{FP}{FP + TN}. \quad (4)$$

The F_1-score (Equation (5)) measures the harmonic mean of precision and recall, which serves as a derived effectiveness measurement:

$$F_1 - Score = 2 \times \frac{Precision * Recall}{Precision + Recall}. \quad (5)$$

In our experiments, based on the flow statistics in the datasets and number of labeled packets in data pre-processing, we can calculate the number of packets (including benign packets) for training and number of real attack packets in testing, as summarized in Tables 5–8 below. Due to a large amount of input data, we set the mini-batch size value at 100 and the data training with less than 200 epochs.

The performance results of our proposal for each kind of malicious traffic with 10-fold testing and 10-fold validation are shown in Tables 9 and 10, respectively. From these two tables, we can observe that our method can perform the classification with nearly 100% accuracy and precision in major cases of the security attacks, which outperforms compared to the prior works, e.g., [6,19,20]. However, since the characteristics of traffic types and attacks are unique, the performance measurements are also different in each scenario and heavily relied on the training dataset. From our experimental experience, we notice that, besides tuning the learning model and its parameters (e.g., learning rate), word-embedding and attack representation samples in datasets play a critical role in the improvement of the performance.

Table 5. Train/Test/Validate traffic proportion in the ISCX-IDS-2012 dataset.

		All	Train/Test	Validation
6/12	Benign	5,947,337	26,374	average:
	Attack	26,374	26,643	4148/60 s
6/13	Benign	3,925,130	100,000	average:
	Attack	1,838,019	100,000	1209 /60 s
6/14	Benign	8,687,942	100,000	average:
	Attack	960,711	100,000	5947 /60 s
6/15	Benign	17,551,503	100,000	average:
	Attack	17,431,539	100,000	12,746 /60 s
6/16	Benign	17,260,920	50,000	average:
	Attack	49,764	49,764	7580 /60 s

Table 6. Train/Test/Validate traffic proportion in the Mirai-RGU dataset.

Attack Type	All	Train /Test	Validation
syn	1,526,926	728,000	1,526,926 /10 s
ack	5,390,837	728,000	5,390,770 /50 s
http	744,991	728,000	3152 /60 s
udp	4,567,726	728,000	4,567,659 /58 s

Table 7. Train/Test/Validate traffic proportion in the USTC-TFC-2016 dataset.

Benign				Malware			
Type	**All**	**Train/Test**	**Validate (avg/60 s)**	**Type**	**All**	**Train/Test**	**Validate (avg./60 s)**
Facetime	6000	6000	6000	Tinba	22,000	22,000	729
Skype	12,000	12,000	12,000	Zeus	93,141	93,141	105
BitTorrent	15,000	15,000	15,000	Shifu	500,000	100,000	3
Gmail	25,000	25,000	25,000	Neris	499,218	100,000	896
Outlook	15,000	15,000	15,000	Cridex	461,548	100,000	34
World Of Warcraft	140,000	100,000	140,000	Nsis-ay	352,266	100,000	5617
MySQL	200,000	100,000	200,000	Geodo	250,000	100,000	12
FTP	360,000	100,000	360,000	Miuref	88,560	88,560	7
SMB	925,453	200,000	925,453	Virut	440,625	100,000	858
Weibo	1,210,060	100,000	1,210,060	Total	2,707,358	803,701	-
Total	2,908,513	1,058,000	-				-

Table 8. Train/Test/Validate traffic proportion in our dataset.

File	No# of Pcap	All (Benign, Attack)	Train/Test	Validate(avg./60 s)
Capture_2	3 (N)	(129,178, 0)	100,000	14,100
Capture_3	1 (M)	(54,641, 393,325)	100,000	9795
Capture_4	6 (N, M)	(268,461, 518,105)	100,000	10,010
Capture_5	1 (N, M)	(67,239, 519,376)	100,000	9780
Capture_6	1 (N, M)	(66,989, 519,609)	100,000	9781
Capture_7	1 (N, M)	(67,061, 519,400)	100,000	9783
Capture_8	1 (N, M, S)	(66,801, 981,651)	100,000	9996
Capture_9	12 (N, M, G)	(72,204, 1,373,042)	100,000	10,004
Capture_10	7 (N, M, GT)	(70,457, 969,937)	100,000	9784

Table 9. Performance evaluation results of the proposed solution on the four datasets with 10-fold testing.

Dataset	Accuracy (%)	Precision (%)	Recall (%)	F-Score (%)	FPR (%)
ISCX-IDS-2012	99.99	99.98	99.99	99.99	7.46×10^{-7}
USTC-TFC-2016	99.99	100	99.99	99.99	1.1×10^{-7}
Mirai-RGU	100	100	100	100	0
Mirai-CCU	99.46	99.63	99.38	99.51	0.026

Table 10. Performance evaluation results of the proposed solution on the four datasets with 10-fold validation.

Dataset	Accuracy (%)	Precision (%)	Recall (%)	F-Score (%)	FPR (%)
ISCX-IDS-2012	99.97	100	99.97	99.98	0
USTC-TFC-2016	99.88	99.99	99.86	99.93	0.02
Mirai-RGU	99.98	99.99	99.95	99.97	0
Mirai-CCU	99.36	99.49	99.27	99.38	0.031

One might question the aforementioned results of our proposed DL-based framework. That is, intuitively, the proposed framework can classify malicious packets because it obtains the features of malicious packets from the training dataset. Thus, we perform a validation test in which the Mirai-RGU dataset is used for training and packets from Mirai-CCU dataset are used for validation.

Our experimental results show that, even though the training and validation stages use different datasets, the proposed framework is still able to detect malicious packets with very high accuracy. The performance results are shown in Table 11.

Table 11. Performance evaluation results of the Mirai-CCU dataset validation on the Mirai-RGU training model.

	Mirai-RGU				
Dataset	**Accuracy (%)**	**Precision (%)**	**Recall (%)**	**F-Score (%)**	**FPR (%)**
Mirai-CCU	97.22	96.25	98.73	97.5	0.36

4.1. Time Efficiency

Time efficiency is also an important metric measurement in our experiment. The time efficiency involves two cases: training time and detection time. The training time is defined by the total time to complete the training on the selected dataset. This time depends on the data-preprocessing, the training model (number of layers and dims) and server capacity. Therefore, it would not be fair to draw comparisons within this respect, due to differences in the hardware and software used. Through our evaluation, due to considering at the deeper level (the packet level), the amount of data (after transferring from packet information) used for training in our method is expected to be higher than the flow-based approaches. As a result, the time required for training will be higher. The time for offline training, i.e., ours, could be up to 17 h with USTC-TFC2016 at 200 epochs, in order to obtain the detection performance results as mentioned above. The bigger dataset it is, the more training time it costs. However, our method has advantages of detection time. The detection time in this study is defined as the total time (elapsed time) to perform the classification on a given testing sample, i.e., the testing dataset. Given a testing file 108 MB and our server capacity (mentioned above), our system only takes fewer than 2 s to complete the classification, no matter what type of traffic is. This achievement emboldens our approach for online monitoring since we believe that this speed can satisfy most on-demand applications and in medium networks. For the core networks, we may need more efforts to integrate this system for potential deployment, since the network speed at such nodes can be up to hundreds of gigabits per second. Note that the system at the core networks can be equipped with the much more powerful servers.

4.2. Discussion

There is a trade-off between the word size and the classification performance, i.e., accuracy and detection time. Reducing the word size can potentially increase the detection time, and training time as well, due to more words involved, longer sentence created and dimension extended. For accuracy, the performance can vary and depend on the considered attack type. Apparently, we can't scale the size too small. In our evaluation, we set the word size to two bytes as aforementioned. However, there is promising research to propose an algorithm to auto-calculate this value properly on the deployed environment, even that for specific attacks.

In addition, the approach to consider the classification at the packet level can open the door to accelerating the detection since we can schedule the packets for parallel processing. However, the explicit shortcoming is to increase the training time and resource usage (e.g., memory) due to a large number of parameters and data size putting to the training model. Balancing the factor of acceptable training time but still gaining the best classification performance is a non-trivial task and a potential research direction.

Finally, the DL-based classification approach is highly susceptible to the data poisoning attack due to its dependence on the training data. Thus far, we have found few attack models targeting the evasion of the deep learning-based malicious classification systems, including ours. However, this can be soon changed when the popularity of deep learning will attract more attackers to exploit its vulnerabilities for hacking or monetization. Generating/preventing adversary models against deep learning is thus one of the most interesting and promising security research topics.

5. Conclusions

In this paper, we have proposed a novel packet-based malicious traffic classification framework by using the word embedding and Long Short-Term Memory (LSTM) network model. Our evaluation results show that the performance of ours is competitive with the prior work on classifying flows into benign or malicious, but with much lower flow pre-processing time. In other words, the major advantage of the proposed framework is that it does not require pre-process packets into flows, thus boosting the detection acceleration. We believe that our first attempt on using a packet-based malicious classification approach can inspire the research community to consider the optimization methods and exploit the advantages of deep learning to build effective IDSs without suffering significant detection delay.

Author Contributions: Conceptualization, R.-H.H. and M.-C.P.; Methodology, R.-H.H. and Y.-L.C.; Data curation, Y.-L.C.; Formal analysis, R.-H.H. and Y.-L.C.; Investigation, R.-H.H., M.-C.P., V.-L.N. and Y.-L.C.; Resources, R.-H.H.; Software, Y.-L.C.; Validation, R.-H.H., V.-L.N. and Y.-L.C.; Visualization, V.-L.N. and Y.-L.C.; Writing—original draft preparation, Y.-L.C.; Writing—review & editing, R.-H.H., V.-L.N. and Y.-L.C.; Supervision, R.-H.H.; Project administration, R.-H.H.; Funding acquisition, R.-H.H.

Funding: This research was supported in part by the Ministry of Science and Technology of Taiwan, ROC, under Grants MOST 107-2218-E-194-014.

Conflicts of Interest: The authors declare no conflict of interest.

References

1. Javaid, A.; Niyaz, Q.; Sun, W.; Alam, M. A deep learning approach for network intrusion detection system. In Proceedings of the 9th EAI International Conference on Bio-Inspired Information and Communications Technologies, New York, NY, USA, 3–5 December 2016; pp. 21–26.
2. Meidan, Y.; Bohadana, M.; Mathov, Y.; Mirsky, Y.; Shabtai, A.; Breitenbacher, D. N-BaIoT—Network-Based Detection of IoT Botnet Attacks Using Deep Autoencoders. *IEEE Pervasive Comput.* **2018**, *17*, 11–22. [CrossRef]
3. Wang, W.; Zhu, M.; Zeng, X.; Ye, X.; Sheng, Y. Malware Traffic Classification Using Convolutional Neural Networks for Representation Learning. In Proceedings of the 31st International Conference on Information Networking, Da Nang, Vietnam, 11–13 January 2017; pp. 712–717. [CrossRef]
4. Chen, Y.C.; Li, Y.J.; Tseng, A.; Lin, T. Deep Learning for Malicious Flow Detection. In Proceedings of the IEEE 28th Annual International Symposium on Personal, Indoor, and Mobile Radio Communications, Montreal, QC, Canada, 8–13 October 2017.
5. Yuan, X.; Li, C.; Li, X. DeepDefense: Identifying DDoS Attack via Deep Learning. In Proceedings of the IEEE International Conference on Smart Computing, Hong Kong, China, 29–31 May 2017. [CrossRef]
6. Yin, C.; Zhu, Y.; Fei, J.; He, X. A deep learning approach for intrusion detection using recurrent neural networks. *IEEE Access* **2017**, *5*, 21954–21961. [CrossRef]
7. Li, C.; Wang, J.; Ye, X. Using a Recurrent Neural Network and Restricted Boltzmann Machines for Malicious Traffic Detection. *NeuroQuantology* **2018**, *6*, 21954–21961. [CrossRef]
8. Hwang, R.H.; Peng, M.C.; Huang, C.W. Detecting IoT Malicious Traffic based on Autoencoder and Convolutional Neural Network. *IEEE Globecom Conf.* **2019**, submitted.
9. Radford, B.J.; Apolonio, L.M.; Trias, A.J.; Simpson, J.A. Network Traffic Anomaly Detection Using Recurrent Neural Networks. *arXiv* **2018**, arXiv:1803.10769.
10. Cui, J.; Long, J.; Min, E.; Mao, Y. WEDL-NIDS: Improving Network Intrusion Detection Using Word Embedding-Based Deep Learning Method. In Proceedings of the International Conference on Modeling Decisions for Artificial Intelligence, Palma de Mallorce, Spain, 15–18 October 2018; pp. 283–295.

11. Kim, M.S.; Kong, H.J.; Hong, S.C.; Chung, S.H.; Hong, J.W. A flow-based method for abnormal network traffic detection. In Proceedings of the 2004 IEEE/IFIP Network Operations and Management Symposium, Seoul, Korea, 23 April 2004; pp. 599–612.
12. Fu, R.; Zhang, Z.; Li, L. Using LSTM and GRU neural network methods for traffic flow prediction. In Proceedings of the 2016 31st Youth Academic Annual Conference of Chinese Association of Automation, Wuhan, China, 11–13 November 2016; pp. 324–328.
13. Goldberg, Y.; Levy, O. word2vec Explained: Deriving Mikolov et al.'s negative-sampling word-embedding method. *arXiv* **2013**, arXiv:1402.3722.
14. Pennington, J.; Socher, R.; Manning, C.D. GloVe: Global vectors for word representation. In Proceedings of the 2014 Conference on Empirical Methods in Natural Language Processing, Doha, Qatar, 25–29 October 2014; pp. 1532–1543.
15. Ronen, R.; Radu, M.; Feuerstein, C.; Yom-Tov, E.; Ahmadi, M. Microsoft Malware Classification Challenge. 2018. Available online: https://arxiv.org/pdf/1802.10135.pdf (accessed on 29 May 2019).
16. McDermott, C.D.; Majdani, F.; Petrovski, A.V. Botnet Detection in the Internet of Things using Deep Learning Approaches. In Proceedings of the 2018 International Joint Conference on Neural Networks (IJCNN), Rio de Janeiro, Brazil, 8–13 July 2018. [CrossRef]
17. Shiravi, A.; Shiravi, H.; Tavallaee, M.; Ghorbani, A. Toward developing a systematic approach to generate benchmark datasets for intrusion detection. *Comput. Secur.* **2012**, *31*, 357–374. [CrossRef]
18. Mikolov, T.; tau Yih, W.; Zweig, G. Linguistic Regularities in Continuous Space Word Representations. In Proceedings of the 2013 Conference of the North American Chapter of the Association for Computational Linguistics: Human Language Technologies, Atlanta, GA, USA, 9–14 June 2013; pp. 746–751.
19. Shone, N.; Ngoc, T.N.; Phai, V.D.; Shi, Q. A Deep Learning Approach to Network Intrusion Detection. *IEEE Trans. Emerg. Top. Comput. Intell.* **2018**, *2*, 41–50. [CrossRef]
20. Wang, W.; Sheng, Y.; Wang, J.; Zeng, X.; Ye, X.; Huang, Y.; Zhu, M. HAST-IDS: Learning Hierarchical Spatial-Temporal Features Using Deep Neural Networks to Improve Intrusion Detection. *IEEE Access* **2018**, *6*, 1792–1806. [CrossRef]

applied sciences

MDPI

Article

Hierarchical Guided-Image-Filtering for Efficient Stereo Matching

Chengtao Zhu [1] and Yau-Zen Chang [2,3,*]

1 School of Microelectronics, Tianjin University, Tianjin 300072, China
2 Department of Mechanical Engineering, Chang Gung University, Taoyuan 33302, Taiwan
3 Department of Neurosurgery, Chang Gung Memorial Hospital, Taoyuan 33305, Taiwan
* Correspondence: zen@mail.cgu.edu.tw; Tel.: +886-3-2118800 (ext. 5341)

Received: 25 June 2019; Accepted: 30 July 2019; Published: 1 August 2019

Featured Application: Potential applications of the work include autonomous navigation, 3D reconstruction, and vision-based object handling.

Abstract: Stereo matching is complicated by the uneven distribution of textures on the image pairs. We address this problem by applying the edge-preserving guided-Image-filtering (GIF) at different resolutions. In contrast to most multi-scale stereo matching algorithms, parameters of the proposed hierarchical GIF model are in an innovative weighted-combination scheme to generate an improved matching cost volume. Our method draws its strength from exploiting texture in various resolution levels and performing an effective mixture of the derived parameters. This novel approach advances our recently proposed algorithm, the pervasive guided-image-filtering scheme, by equipping it with hierarchical filtering modules, leading to disparity images with more details. The approach ensures as many different-scale patterns as possible to be involved in the cost aggregation and hence improves matching accuracy. The experimental results show that the proposed scheme achieves the best matching accuracy when compared with six well-recognized cutting-edge algorithms using version 3 of the Middlebury stereo evaluation data sets.

Keywords: stereo matching; cost aggregation; image filtering; binocular stereo vision

1. Introduction

Stereo vision aims at providing rich distance information of the captured scenes via image pairs. This is normally accomplished by matching algorithms to generate dense disparity maps. The maps can be transformed into three-dimensional information of the scene by the principle of triangulation with many potential applications, such as autonomous navigation, 3D reconstruction, and vision-based object handling.

Although the stereo matching problem has been under extensive research for decades, it is still difficult to obtain accurate matching under ill-posed conditions such as texture-less regions, repeated patterns, occlusion areas, and reflective surfaces. The current stereo matching algorithms can mainly be divided into two categories: The conventional matching algorithms [1] and the deep-learning-based stereo matching approaches.

The stereo matching algorithm based on deep learning regards the process of deriving the disparity map as a classification problem or a regression problem. For instance, Zbontar [2] used the convolutional neural networks (CNNs) to estimate the similarity of image blocks and uses the measures as the matching cost in the traditional stereo matching algorithm. Similarly, Nahar [3] proposed unsupervised pre-trained networks to estimate hierarchical features and combine them with a pixel-based intensity matching cost in a global energy minimization framework for dense disparity estimation. By combining a disparity estimation network with a CNN that was trained by a synthetically generated dataset, Mayer [4] demonstrated the effectiveness of deep learning in stereo matching. Pang [5] proposed a cascaded CNN architecture that is composed of two stages: The first stage advances the work of Reference [4] by equipping it with extra up-convolution modules, while the second stage generates residual signals across multiple scales. The summation of the outputs from the two stages gives the final disparity. Kendall [6] used deep unary features to compute a stereo matching cost volume. In this approach, disparity values are regressed for aggregation from the cost volume using 3D convolutions.

Another method to implement deep learning-based stereo matching is to use the networks to exploit context information. For example, Chang [7] developed a spatial pyramid pooling module to the aggregate context in different scales to form a cost volume. The cost volume is regularized by a stacked network to further improve the utilization of global context information. Besides, Williem [8] used the deep learning technique for the cost volume aggregation based on self-guided filtering.

Deep learning-based methods are promising as they can apply the high-level object detection as a guideline for within-object matching. However, most of the current schemes use supervised learning methods that assume the true disparity is known in advance. This assumption is impractical for many applications [9]. Moreover, these approaches might be invalid for an unknown environment and cannot be well transplanted to robotic and embedded systems [10].

The conventional stereo matching approaches are classified as global or local according to the construction of an objective function that rates the degree of match between an image pair [1]. The objective function of the global methods consists of a data term (the measurement part) and a regularization term (the penalty part). The data term designates the similarity between aggregated matching costs of pixels on the images, and the regularization term is included to provide constraints from neighboring pixels. Belief propagation [11] and dynamic programming [12,13] are the major global methods. However, global approaches need a lot of computing resource and are generally not suitable for real-time applications.

In contrast to the global approaches, the objective function of the local methods contains only the measurement part. The local methods generally perform the stereo matching in four stages [1]: (1) The calculation of the preliminary matching cost, (2) aggregation of the cost over support windows, (3) estimation of the disparity, and (4) refining the disparity. Among them, the cost aggregation step is usually transformed into an image filtering procedure of the matching cost and the disparity maps are obtained by the winner-takes-all method [1]. The local methods require less computation and are popular for fast disparity calculations.

Cost aggregation is crucial for matching performance in the local algorithms. Bilateral filtering [14] is among the early approaches that led to the increase of computational complexity with the increase of support-window size. Later, tree filtering [15], domain transformation [16], recursive edge-aware filter [17], and full-image guided filtering [18] were proposed to decouple computational complexity with the support window size. However, these approaches all suffer from the weight-decay problem when there is a significant intensity difference between neighboring pixels. This behavior deteriorates information propagation and impairs the resulting matching performance.

Hosni [19] suggested treating the generation of disparity as a labeling problem, which is implemented through the steps of constructing a cost volume, cost volume filtering, followed by winner-takes-all label selection. Along this line, the guided-image-filter (GIF) [20] substantially involves cost volume filtering because it can generate clear edge profiles free from the gradient-reversal artifacts. Later, Li [21] introduced an edge-aware weighting, denoted as the weighted guided-image-filtering (WGIF), to improve GIF. Kou [22] proposed a gradient-domain guided-image-filter (GDGIF) to reduce halo artifacts by incorporating an explicit first-order edge-aware constraint. Nevertheless, due to the lack of pixel information outside the fixed window, the implementation of WGIF by Hong [23] results in restricted performance.

To remove the fixed-window limitation, approaches with adaptive guided filters were proposed [24]. However, information outside the support windows is still missing. In a recent paper [25], we introduced weights that take both distance and intensity differences into account to extend the scheme of GIF. We called our approach the pervasive guided-image-filtering, denoted as PGIF [25], which exploits the whole image for aggregation.

Also, recent years have seen the development of coarse-to-fine (CTF) strategies to enhance the stereo matching accuracy. For instance, Hu [26] proposed to reduce the search space of local stereo matching by introducing a candidate set of neighbor pixels. Tang [27] introduced a multi-scale pixel feature vector to provide effective matching of radiometric differences. Advanced techniques to find improved disparity range was provided in the work of Li [28] by recursive multi-scale decomposition. These methods assume the existence of disparity consistency. In contrast to these approaches, the matching cost integration method of Zhang [29] resulted in superior matching, where cost aggregation is formulated to enforce the consistency of the cost volume among the neighboring scales.

Inspired by the multiscale scheme of Zhang [29], we extend the pervasive guided-image-filtering, PGIF [25], to exploit the cross-scale features in the cost volume. In our approach, the consistency is imposed on the GIF parameters, rather than the cost volume of Zhang [29], in the neighboring-scale direction. We call it hierarchical guided-image-filtering, denoted as HGIF.

The main contribution of this paper can be summarized as:

(1) We created an innovative aggregation approach that efficiently combines the model parameters of PGIF [25] to allow the features of the image pairs in different resolutions to be considered;
(2) The scheme is unique in its parameter-based aggregation, rather than the cost-volume-based approaches in the current literature, allowing efficient calculation with superior performance;
(3) The proposed scheme outperforms most of the state-of-art algorithms in terms of disparity accuracy even without the refinement procedure.

2. Proposed Method

2.1. The Cost Aggregation Based on the Pervasive Guided-Image-Filtering (PGIF)

This section presents the basic procedure of the pervasive guided-image-filtering (PGIF) [25] scheme. It serves as the definitions of some of the variables and algorithms for the proposed scheme to be shown in the next section.

In the GIF-based stereo matching algorithms, such as WGIF [21], GDGIF [22], and PGIF [25] are discussed in the last section. The aggregated cost, $\overline{C}_{d,0}(p)$ is a linear function of the guidance image, I_G at a local disparity patch such that:

$$\overline{C}_{d,0}(p) = a_{d,0}(p) \cdot I_G(p) + b_{d,0}(p), \tag{1}$$

where $p = (x, y)$ is the location of the central pixel and d indicates the disparity. Note that $\overline{C}_{d,0}(p)$ have three dimensions (x, y, d) and is often denoted as the aggregated cost volume. Likewise, $a_{d,0}(p)$ and $b_{d,0}(p)$ can also be called the parameter volumes of the model. The second subscript, 0, indicates that these values are related to the original resolution. For the following stereo matching operations, we regard the left image as the reference image, such that $I_G = I_{L,0}$, and the right image as the target image.

The disparity map, $D(p)$ is composed of the disparity value corresponding to the minimum aggregated cost at each location p:

$$D(p) = \underset{d}{\operatorname{argmin}} \overline{C}_{d,0}(p). \tag{2}$$

In order to solve for $a_{d,0}(p)$ and $b_{d,0}(p)$, we proposed an objective function in Reference [25], which includes a weighted sum of the squared difference between the linear model and a primary matching cost, denoted as $C_{d,0}(q)$, and a regularization term:

$$E_{d,0}(p) = \sum_{q \in \Omega_0(p)} \omega_0(p,q) \cdot \{[a_{d,0}(p) \cdot I_{L,0}(q) + b_{d,0}(p) - C_{d,0}(q)]^2 + \varepsilon(p) \cdot [a_{d,0}(p)]^2\} \tag{3}$$

where Ω_0 is the support window for a pixel located at p, $\omega_0(p,q)$ is the weight that reflects the contribution of a pixel located at q, and $\varepsilon(p)$ is the regularization factor to limit the magnitude of $a_{d,0}(p)$. Because of an efficient iterative computation scheme, to be presented below, the support window is the whole image.

The primary matching cost, $C_{d,0}(q)$, represents the degree of match between a pixel located at q of the left image and a pixel located at $q + d$ of the right image. This cost is smaller for a higher degree of match. We employ a truncated version of the absolute gradient difference to define the cost:

$$C_{d,0}(q) = \min(\left|\nabla_x I_{L,0}(q) - \nabla_x I_{R,0}(q+d)\right|, \tau) + \min(\left|\nabla_y I_{L,0}(q) - \nabla_y I_{R,0}(q+d)\right|, \tau) \tag{4}$$

where $I_{L,0}$ and $I_{R,0}$ are the left and right images of the stereo pair in the original resolution, ∇_x and ∇_y are the horizontal and vertical gradients, respectively, and τ is the truncation threshold, normally assigned as 2. The use of the threshold helps to reduce the mismatch in obscured or noisy regions.

The optimum values of the parameters in Equation (3), represented as $a^*_{d,0}(p)$ and $b^*_{d,0}(p)$, are obtained by minimizing the objective function Equation (3):

$$\begin{aligned} \frac{\partial E_{d,0}(p)}{\partial a_{d,0}(p)} &= 0 \\ \frac{\partial E_{d,0}(p)}{\partial b_{d,0}(p)} &= 0 \end{aligned} \tag{5}$$

We have:

$$\begin{aligned} a^*_{d,0}(p) &= \frac{\dfrac{\sum_{q\in\Omega_0(p)} [\omega_0(p,q)\cdot I_{L,0}(q)\cdot C_{d,0}(q)]}{\sum_{q\in\Omega_0(p)} \omega_0(p,q)} - \dfrac{\sum_{q\in\Omega_0(p)} [\omega_0(p,q)\cdot I_{L,0}(q)]}{\sum_{q\in\Omega_0(p)} \omega_0(p,q)} \cdot \dfrac{\sum_{q\in\Omega_0(p)} [\omega_0(p,q)\cdot C_{d,0}(q)]}{\sum_{q\in\Omega_0(p)} \omega_0(p,q)}}{\dfrac{\sum_{q\in\Omega_0(p)} [\omega_0(p,q)\cdot I_{L,0}(q)\cdot I_{L,0}(q)]}{\sum_{q\in\Omega_0(p)} \omega_0(p,q)} - \left\{\dfrac{\sum_{q\in\Omega_0(p)} [\omega_0(p,q)\cdot I_{L,0}(q)]}{\sum_{q\in\Omega_0(p)} \omega_0(p,q)}\right\}^2 + \varepsilon(p)} \\ b^*_{d,0}(p) &= \frac{\sum_{q\in\Omega_0(p)} [\omega_0(p,q)\cdot C_{d,0}(q)]}{\sum_{q\in\Omega_0(p)} \omega_0(p,q)} - a^*_{d,0}(p) \cdot \frac{\sum_{q\in\Omega_0(p)} [\omega_0(p,q)\cdot I_{L,0}(q)]}{\sum_{q\in\Omega_0(p)} \omega_0(p,q)} \end{aligned} \tag{6}$$

In PGIF [25], the weight $\omega_0(p,q)$ is decomposed into horizontal and vertical weighting factors, $\overline{W}^H_{i,x,0}$ and $\overline{W}^V_{j,y,0}$:

$$\omega_0(p,q) = \omega_0(x,y,i,j) = \overline{W}^H_{i,x,0} \cdot \overline{W}^V_{j,y,0} \tag{7}$$

With the convention that $q = (i,j)$ and $p = (x,y)$, the weighting factors, $\overline{W}^H_{i,x,0}$ and $\overline{W}^V_{j,y,0}$, can be recursively calculated as:

$$\begin{aligned} \overline{W}^H_{i,x,0} &= \begin{cases} \prod_{u=\min(i,x)+1}^{\max(i,x)} \exp\left(\frac{f(|I_{L,0}(u,j)-I_{L,0}(u-1,j)|)}{-\beta}\right), & i \neq x \\ 1, & i = x \end{cases} \\ \overline{W}^V_{j,y,0} &= \begin{cases} \prod_{u=\min(j,y)+1}^{\max(j,y)} \exp\left(\frac{f(|I_{L,0}(x,u)-I_{L,0}(x,u-1)|)}{-\beta}\right), & j \neq y \\ 1, & j = y \end{cases} \\ &\text{with } f(\xi) = \begin{cases} 0, & \text{if } \xi = 0 \\ 1, & \text{if } \xi > 0 \end{cases} \end{aligned} \tag{8}$$

where the parameter β is a constant factor. By this scheme, the weight $\omega_0(p,q)$ depends on both the spatial and intensity differences, and the contribution of any pixel located at q to the pixel located at p can be effectively involved. Besides, the introduction of function f alleviates the effects of abrupt change in intensity, enhancing the immunity to noise caused by the recursive calculation.

2.2. Stereo Matching Based on Hierarchical Guided-Image-Filtering (HGIF)

A major shortage of approaches using the scheme starting from an energy function of Equation (3), including WGIF [21], GDGIF [22], and PGIF [25], is the restriction of the cost metrics to local regions of the support window $\Omega_0(p)$. Even though PGIF [25] exploits the whole image for matching, the effects of patterns far from the pixel under aggregation decay with distance.

The proposed scheme begins with a down-sampling of the original image pair, $I_{L,0}$ and $I_{R,0}$, where the left image is selected as the reference image and the right image as the target image. As the images are down-sampled by a factor of 2, we denote the resultant image pairs as $I_{L,z}$ and $I_{R,z}$ for $z \in \{0, 1, \ldots, K\}$ with K being the roughest level.

Firstly, the optimal parameters, $a^*_{d,z}(q)$ and $b^*_{d,z}(q)$, of each image pair, $I_{L,z}$ and $I_{R,z}$, are calculated using Equation (6). We define a set of unknown parameters, $\hat{a}_{d,z}(p)$ and $\hat{b}_{d,z}(p)$, as aggregated parameters to be found. In order to aggregate the cross-scale patterns, we create an objective function $F_d(p)$ that is the weighted mean-square error between the optimized parameters, $a^*_{d,z}(q)$ and $b^*_{d,z}(q)$, and the unknown parameters, $\hat{a}_{d,z}(p)$ and $\hat{b}_{d,z}(p)$, throughout each scale:

$$\begin{aligned} F_d(p) = &\sum_{z=0}^{K} \frac{\sum_{q \in I_{L,z}(p)} \omega_z(p,q) \cdot [a^*_{d,z}(q) - \hat{a}_{d,z}(p)]^2}{\sum_{q \in I_{L,z}(p)} \omega_z(p,q)} + \sum_{z=0}^{K} \frac{\sum_{q \in I_{L,z}(p)} \omega_z(p,q) \cdot [b^*_{d,z}(q) - \hat{b}_{d,z}(p)]^2}{\sum_{q \in I_{L,z}(p)} \omega_z(p,q)} \\ &+ \sum_{z=1}^{K} \gamma^z \cdot \left([\hat{a}_{d,z}(p) - \hat{a}_{d,z-1}(p)]^2 + [\hat{b}_{d,z}(p) - \hat{b}_{d,z-1}(p)]^2 \right), \end{aligned} \tag{9}$$

where the weight $\omega_z(p,q)$ is defined similarly to Equations (7) and (8) with the subscript 0 replaced by z. This objective function inherits the weighting scheme of Reference [25] that not only reflects the effects of distance and intensity difference, but also applies constraints on the GIF parameters in the scale direction.

In Equation (9), the positive constant γ is a constraint factor on the squared difference between the parameters $\hat{a}_{d,z}(p)$ and $\hat{a}_{d,z-1}(p)$, and between $\hat{b}_{d,z}(p)$ and $\hat{b}_{d,z-1}(p)$, respectively. The constraint weights, γ^z for $z \in \{0, 1, \ldots, K\}$, are larger between layers that are far from the original image, layer 0, if $\gamma > 1$. This trend is reversed for $\gamma < 1$.

There are totally $2(K+1)$ unknown parameters in Equation (9), namely $\hat{a}_{d,0}(p), \hat{b}_{d,0}(p), \ldots, \hat{a}_{d,K}(p)$, and $\hat{b}_{d,K}(p)$. They can be obtained by setting the partial derivatives of Equation (9) with respect to these parameters to zero:

$$\begin{cases} \dfrac{\partial F_d(p)}{\partial \hat{a}_{d,z}(p)} = 0 \\ \dfrac{\partial F_d(p)}{\partial \hat{b}_{d,z}(p)} = 0 \end{cases}, \quad z \in \{0,\ 1,\ 2,\ \ldots, K\} \tag{10}$$

Equation (10) can be simplified as:

$$\begin{cases} (1+\gamma) \cdot \hat{a}_{d,0}(p) - \gamma \cdot \hat{a}_{d,1}(p) = g_z(a^*_{d,z}(p)), & \text{for } z = 0 \\ -\gamma^z \cdot \hat{a}_{d,z-1}(p) + (1+\gamma^z+\gamma^{z+1}) \cdot \hat{a}_{d,z}(p) - \gamma^{z+1} \cdot \hat{a}_{d,z+1}(p) = g_z(a^*_{d,z}(p)), & \text{for } z \in \{1,\ 2, \ldots, K-1\} \\ -\gamma^K \cdot \hat{a}_{d,K-1}(p) + (1+\gamma^K) \cdot \hat{a}_{d,K}(p) = g_z(a^*_{d,z}(p)), & \text{for } z = K \end{cases} \tag{11}$$

and:

$$\begin{cases} (1+\gamma) \cdot \hat{b}_{d,0}(p) - \gamma \cdot \hat{b}_{d,1}(p) = g_z(b^*_{d,z}(p)), & \text{for } z = 0 \\ -\gamma^z \cdot \hat{b}_{d,z-1}(p) + (1+\gamma^z+\gamma^{z+1}) \cdot \hat{b}_{d,z}(p) - \gamma^{z+1} \cdot \hat{b}_{d,z+1}(p) = g_z(b^*_{d,z}(p)), & \text{for } z \in \{1,\ 2, \ldots, K-1\} \\ -\gamma^K \cdot \hat{b}_{d,K-1}(p) + (1+\gamma^K) \cdot \hat{b}_{d,K}(p) = g_z(b^*_{d,z}(p)), & \text{for } z = K, \end{cases} \tag{12}$$

where g_z is a nominal average function to the parameters, $a^*_{d,z}(q)$ and $b^*_{d,z}(q)$, with scale-dependent weights $\omega_z(p,q)$:

$$g_z(a^*_{d,z}(p)) = \frac{\sum\limits_{q\in I_{L,z}(p)} \omega_z(p,q)\cdot a^*_{d,z}(q)}{\sum\limits_{q\in I_{L,z}(p)} \omega_z(p,q)}$$
$$g_z(b^*_{d,z}(p)) = \frac{\sum\limits_{q\in I_{L,z}(p)} \omega_z(p,q)\cdot b^*_{d,z}(q)}{\sum\limits_{q\in I_{L,z}(p)} \omega_z(p,q)} \tag{13}$$

The aggregated parameters, $\hat{a}_{d,0}(p)$ and $\hat{b}_{d,0}(p)$, can be solved using the system of linear Equations (11) and (12). Taking the case of $K = 2$ (down-sampled twice) and $\gamma = 1.5$ as an example, Equation (11) can be calculated as:

$$\begin{cases} 2.5\cdot\hat{a}_{d,0}(p) - 1.5\cdot\hat{a}_{d,1}(p) = g_0(a^*_{d,0}(p)) \\ -1.5\cdot\hat{a}_{d,0}(p) + 4.75\cdot\hat{a}_{d,1}(p) - 2.25\cdot\hat{a}_{d,2}(p) = g_1(a^*_{d,1}(p)) \\ -2.25\cdot\hat{a}_{d,1}(p) + 3.25\cdot\hat{a}_{d,2}(p) = g_2(a^*_{d,2}(p)) \end{cases} \tag{14}$$

We have that:

$$\hat{a}_{d,0}(p) = 0.557\cdot g_0(a^*_{d,0}(p)) + 0.262\cdot g_1(a^*_{d,1}(p)) + 0.181\cdot g_2(a^*_{d,2}(p)). \tag{15}$$

Likewise, $\hat{b}_{d,0}(p)$ is a linear combination of the parameters, $b^*_{d,0}(p)$, $b^*_{d,1}(p)$ and $b^*_{d,2}(p)$:

$$\hat{b}_{d,0}(p) = 0.557\cdot g_0(b^*_{d,0}(p)) + 0.262\cdot g_1(b^*_{d,1}(p)) + 0.181\cdot g_2(b^*_{d,2}(p)) \tag{16}$$

Thus, these aggregated parameters, $\hat{a}_{d,0}(p)$ and $\hat{b}_{d,0}(p)$, include the effect of features come from different resolutions. These parameters are three-dimensional with dimensions (x, y, d) where the x and y dimensions are in the image planes and the d dimension is in the disparity direction.

After obtaining the aggregated parameters, the cost volume $\hat{C}_d(p)$ is calculated according to Equation (1):

$$\hat{C}_d(p) = \hat{a}_{d,0}(p)\cdot I_{L,0}(p) + \hat{b}_{d,0}(p) \tag{17}$$

This calculation is conducted using elementwise multiplications and additions in the image plane for each disparity. Finally, the disparity map, $\hat{D}(p)$, can be obtained using the winner-take-all (WTA) procedure of Equation (2):

$$\hat{D}(p) = \underset{d}{\operatorname{argmin}}\hat{C}_d(p). \tag{18}$$

The procedure of our proposed algorithm for stereo matching is depicted in Figure 1, taking $K = 2$ as an example, and summarized in the following steps:

1. Down-sample the image pairs to build a pair of pyramids of images, $I_{L,z}$ and $I_{R,z}$ for $z \in \{0, 1, \ldots, K\}$, with $K + 1$ levels of resolution;

2. Calculate the weight matrix $\omega_z(p,q)$ as a multiplication of two weighting factors, $\overline{W}^H_{i,x,z}$ and $\overline{W}^V_{j,y,z}$ for $z \in \{0, 1, \ldots, K\}$:

$$\omega_z(p,q) = \overline{W}^H_{i,x,z}\cdot\overline{W}^V_{j,y,z} \tag{19}$$

where the weighting factors are recursively calculated for $z \in \{0, 1, \ldots, K\}$ as:

$$\overline{W}^{H}_{i,x,z} = \begin{cases} \prod_{u=\min(i,x)+1}^{\max(i,x)} \exp\left(\frac{f(|I_{L,z}(u,j)-I_{L,z}(u-1,j)|)}{-\beta}\right), & i \neq x \\ 1, & i = x \end{cases}$$
$$\overline{W}^{V}_{j,y,z} = \begin{cases} \prod_{u=\min(j,y)+1}^{\max(j,y)} \exp\left(\frac{f(|I_{L,z}(x,u)-I_{L,z}(x,u-1)|)}{-\beta}\right), & j \neq y \\ 1, & j = y \end{cases} \quad (20)$$
$$\text{with } f(\xi) = \begin{cases} 0, & \text{if } \xi = 0 \\ 1, & \text{if } \xi > 0 \end{cases}$$

3. Generate the primary matching cost volume for each scale where $z \in \{0, 1, \ldots, K\}$:

$$C_{d,z}(q) = \min(|\nabla_x I_{L,z}(q) - \nabla_x I_{R,z}(q+d)|, \tau) + \min(|\nabla_y I_{L,z}(q) - \nabla_y I_{R,z}(q+d)|, \tau). \quad (21)$$

4. Find the optimum parameters for each resolution, where $z \in \{0, 1, \ldots, K\}$:

$$a^{*}_{d,z}(p) = \frac{\frac{\sum_{q \in I_{L,z}(p)} [\omega_z(p,q) \cdot I_{L,z}(q) \cdot C_{d,z}(q)]}{\sum_{q \in I_{L,z}(p)} \omega_z(p,q)} - \frac{\sum_{q \in I_{L,z}(p)} [\omega_z(p,q) \cdot I_{L,z}(q)]}{\sum_{q \in I_{L,z}(p)} \omega_z(p,q)} \cdot \frac{\sum_{q \in I_{L,z}(p)} [\omega_z(p,q) \cdot C_{d,z}(q)]}{\sum_{q \in I_{L,z}(p)} \omega_z(p,q)}}{\frac{\sum_{q \in I_{L,z}(p)} [\omega_z(p,q) \cdot I_{L,z}(q) \cdot I_{L,z}(q)]}{\sum_{q \in I_{L,z}(p)} \omega_z(p,q)} - \left\{\frac{\sum_{q \in I_{L,z}(p)} [\omega_z(p,q) \cdot I_{L,z}(q)]}{\sum_{q \in I_{L,z}(p)} \omega_z(p,q)}\right\}^2 + \varepsilon(p)}$$
$$b^{*}_{d,z}(p) = \frac{\sum_{q \in I_{L,z}(p)} [\omega_z(p,q) \cdot C_{d,z}(q)]}{\sum_{q \in I_{L,z}(p)} \omega_z(p,q)} - a^{*}_{d,z}(p) \cdot \frac{\sum_{q \in I_{L,z}(p)} [\omega_z(p,q) \cdot I_{L,z}(q)]}{\sum_{q \in I_{L,z}(p)} \omega_z(p,q)}. \quad (22)$$

5. Calculate the nominal average function to the parameters, $g_z(a^{*}_{d,z}(p))$ and $g_z(b^{*}_{d,z}(p))$, according to Equation (13) for $z \in \{0, 1, \ldots, K\}$;
6. Solve for $\hat{a}_{d,0}(p)$ and $\hat{b}_{d,0}(p)$ based on the system of linear Equations (11) and (12);
7. Aggregate the cost volume across multiple scales according to Equation (17);
8. Find the disparity map using the WTA procedure of Equation (18).

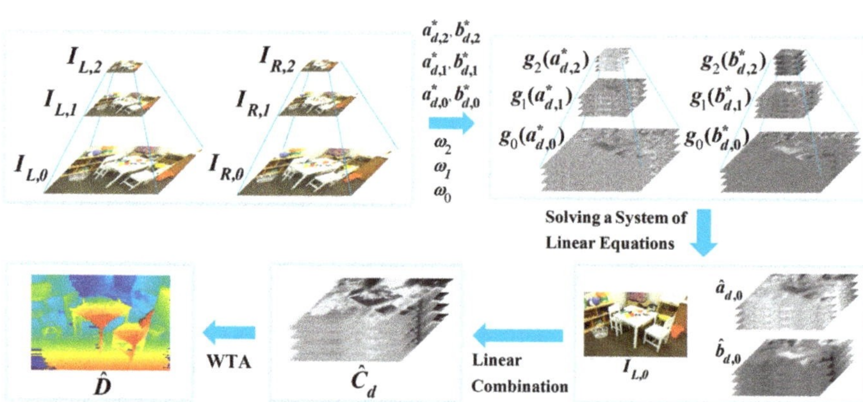

Figure 1. An overview of the proposed stereo matching process with two levels of down-sampling as an example.

As depicted in Figure 1, the parameters $\hat{a}_{d,0}(p)$ and $\hat{b}_{d,0}(p)$ are found by solving the system of linear equations, composed of Equations (11) and (12). As the matrix inverse can be conducted in advance, the calculation can be simplified into matrix multiplication, as demonstrated in Equation (14).

Besides, the cost volume $\hat{C}_d(p)$ is calculated using the linear model of Equation (17). In contrary to the current multi-scale approaches in the literature, such as References [27–29], where the cross-scale aggregation is based on costs themselves, the parameter-based cross-resolution aggregation of the proposed procedure is unique and efficient.

3. Experimental Results

In the proposed scheme, there are two design parameters, β in Equation (20) and γ in Equation (9). According to Equations (19) and (20), we have that larger β will cause $\overline{W}^H_{i,x,z}$ and $\overline{W}^V_{j,y,z}$ to increase, so ω_z is larger and $\hat{D}(p)$ will be smoother. Based on Equation (9), we also have that when γ is larger, the constraint between the scale layers is stronger.

We conducted experiments using the training dataset of the KITTI Vision Benchmark Suite [30] to determine the proper values for these two parameters. The dataset contains 200 image pairs and 200 ground truth disparity maps. The results of the performance of the proposed scheme using the dataset are summarized in Figure 2.

Figure 2a,b display the mean values of the percentage of the erroneous pixels on the disparity maps, denoted as the average error rates, and the standard deviations. Each of the mean values along with the standard deviation was calculated using these 200 image pairs. When $\beta = 2$ and $\gamma = 1.5$ these two parameters achieved their best performance. We fixed these values for all the calculation experiments presented in this section.

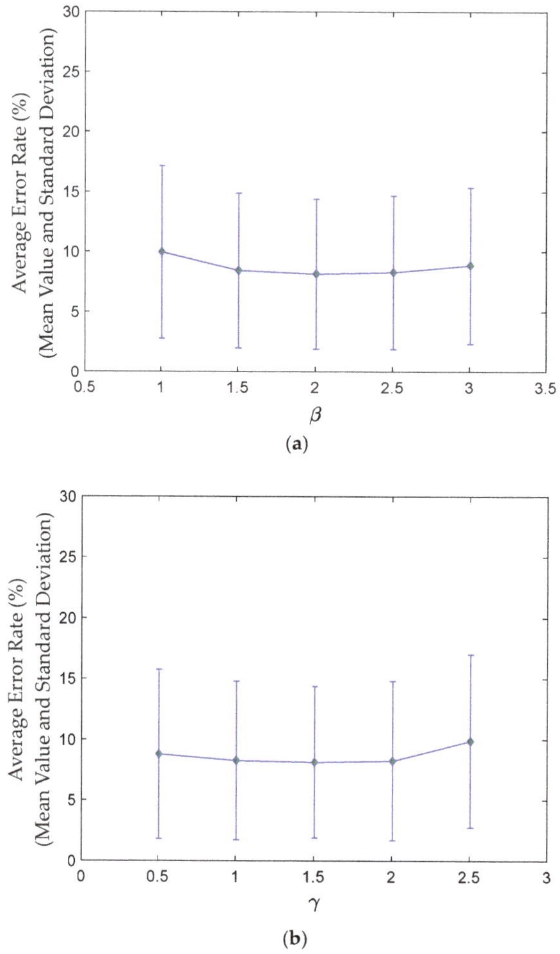

Figure 2. The effect of parameter selection on the performance of the proposed scheme using 200 stereo image pairs of the training dataset in the KITTI Vision Benchmark Suite [30]. The dataset was downloaded from the URL: www.cvlibs.net/datasets/kitti/index.php. Each figure shows the mean values of the error rates and their corresponding standard deviations. (**a**) The average error rate with respect to the parameter β when $\gamma = 1.5$. (**b**) The average error rate with respect to the parameter γ when $\beta = 2$.

To validate the effectiveness of the proposed scheme, extensive comparative experiments were conducted. We studied six state-of-the-art stereo matching algorithms to compare with the proposed scheme:

- The fast cost volume filtering scheme of Reference [19], denoted as FCVF;
- A combination of the cross-scale cost aggregation scheme of Reference [29] and FCVF [19], denoted as CS-FCVF;
- The pervasive guided-image-filter scheme of Reference [25], denoted as PGIF;
- A combination of the cross-scale cost aggregation scheme of Reference [29] and PGIF [25], denoted as CS-PGIF;
- The deep self-guided cost aggregation scheme of Reference [8], denoted as DSG;

- The sparse representation over discriminative dictionary scheme of Reference [13], denoted as SRDD;
- The proposed scheme, which implements a hierarchical guided-image-filter, denoted as HGIF.

We tested these frameworks using the Middlebury (version 3) benchmark stereo database downloaded via the URL: vision.Middlebury.edu/stereo/ [31]. The "trainingQ" image set, which contains 15 stereo pairs, were used for the performance evaluation, starting from "Adirondack" to "Vintage". Of them, only five are shown in Figure 3 due to the space limitation. They are: "Adirondack", "Pipes", "Playroom", "Playtable", and "Shelves". These pictures are with a typical resolution of 720 by 480. These image pairs are down-sized twice, to 360 by 240 and 180 by 120, for example, in the proposed scheme. This arrangement corresponds to $K = 2$.

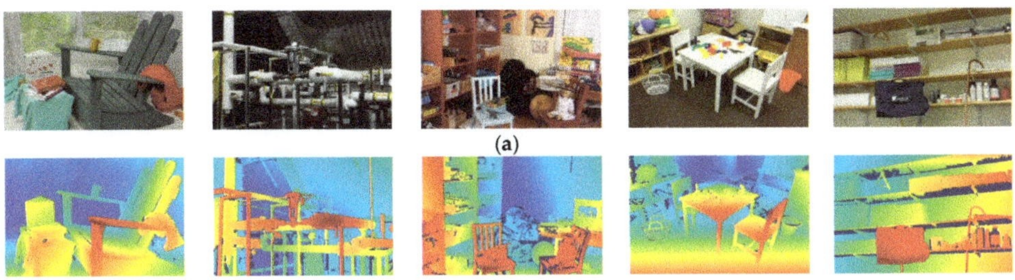

Figure 3. Datasets and their corresponding ground truth disparity maps selected from the experimented data for visual comparison. (**a**) Left images of the image pairs: Adirondack, Pipes, Playroom, Playtable, and Shelves; (**b**) ground truth disparity maps of these images. Image courtesy of the Middlebury (version 3) benchmark stereo database via the URL: vision.middlebury.edu/stereo/ [31].

Besides, the Middlebury defines two measures for evaluating average error rates, including non-occluded (non-occ) and all regions. The weighted average error rate is an official metric of the benchmark in measuring the accuracy of matching by using different weights for different image pairs. These weights are employed to compensate for the variation in the matching difficulty, as remarked on the website. Specifically, the image pairs: "PianoL", "Playroom", "Playtable", "Shelves", and "Vintage" contribute only half of the error rates.

Figure 4 shows the disparity maps obtained by these algorithms. Among them, only the results of SRDD [13] were improved by the disparity refinement procedure. The corresponding error rates in the non-occluded region and the all-region are summarized in Tables 1 and 2, respectively.

Taking a close look of the disparity maps of Figure 4, we have that there was a significant improvement in the matching quality of CS-FCVF [19,29] over FCVF [19], especially in the texture-less regions of the "Playtable" and "Shelves" cases. This improvement was less significant but could also be observed in that of the CS-PGIF [25,29] over PGIF [25]. These improvements were due to the effective cross-scale cost aggregation scheme of Reference [29].

According to the weighted average error rates listed in Tables 1 and 2, DSG [8] performed worst. Also, the matching performance of SRDD [13] was better than CS-FCVF [19,29] but slightly worse than CS-PGIF [25,29]. It is worthy to note that SRDD [13] applies semi-global cost aggregation and post-processing refinement to further improve the matching accuracy. However, the proposed scheme, even without refinement, performed better than SRDD [13] in most of the cases and achieves the smallest weighted average error rates in both the all-region and the non-occluded (non-occ) region.

The experiment was executed in MATLAB 2017b using an Intel Core I5 8300H and 16 GB RAM. Table 3 summarizes the execution time of these algorithms. We have that the multi-scale versions, CS-FCVF [19,29] and CS-PGIF [25,29], required more time for computation than their original versions, FCVF [19] and PGIF [25], respectively, as expected.

Similarly, the proposed algorithm needed to calculate the filtering parameters in multiple scale layers, it ran longer than the FCVF algorithm [19] and the PGIF algorithm [25]. In addition, the CS-FCVF [19,29] and CS-PGIF [25,29] algorithms only incorporated one matching cost parameter, while the proposed algorithm has to compute two parameters, $\hat{a}_{d,0}(p)$ and $\hat{b}_{d,0}(p)$ in Equations (11) and (12). We also have that the proposed algorithm ran slightly longer than CS-FCVF [19,29] and CS-PGIF [25,29]. However, the increased computation time was marginal and still within the same order of magnitude. Moreover, both DSG [8] and SRDD [13] required much more computational resource than the other algorithms due to their deep neural network and semi-global cost aggregation scheme, respectively, as pointed out in Reference [13]. Based on Table 3, we may conclude that the proposed scheme had the best performance when taking both the matching correctness and the calculation efficiency into consideration.

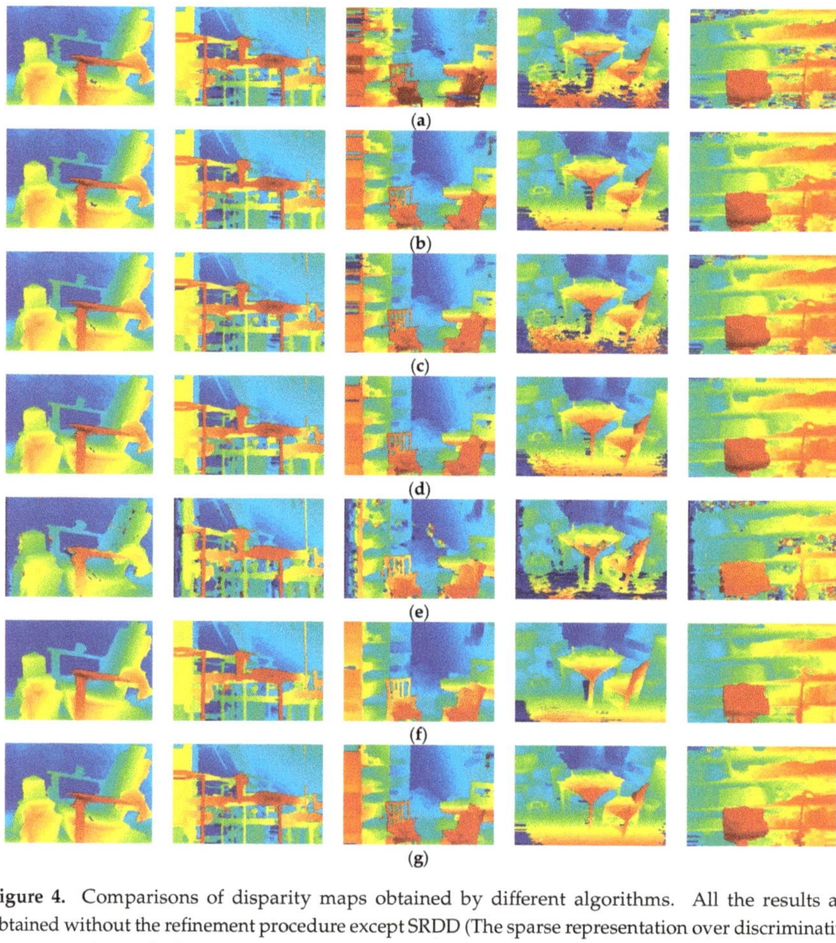

Figure 4. Comparisons of disparity maps obtained by different algorithms. All the results are obtained without the refinement procedure except SRDD (The sparse representation over discriminative dictionary scheme) [13]. (**a**–**g**) are disparity maps generated by: (**a**) FCVF (The fast cost volume filtering scheme) [19], (**b**) CS-FCVF (A combination of the cross-scale cost aggregation scheme [29] and FCVF), (**c**) PGIF (The pervasive guided image filter scheme) [25], (**d**) CS-PGIF (A combination of [29] and PGIF), (**e**) DSG (The deep self-guided cost aggregation scheme) [8], (**f**) SRDD (The sparse representation over discriminative dictionary scheme) [13], and (**g**) the proposed scheme. Among them, the images of (**e**) DSG and (**f**) SRDD are by courtesy of the Middlebury (version 3) benchmark stereo database via the URL: vision.middlebury.edu/stereo/ [31].

Table 1. Comparison of the weighted average error rates in the non-occluded (non-occ) region between seven algorithms (%). All of the results are obtained without the refinement procedure except SRDD [13]. The lowest error records are marked in bold.

Image Sets	FCVF	CS-FCVF	PGIF	CS-PGIF	DSG	SRDD	Proposed
Adirondack	8.78	7.93	6.43	6.30	8.98	6.53	**5.73**
ArtL	12.22	12.20	10.90	**10.78**	13.79	13.17	10.85
Jadeplant	21.81	22.85	20.03	20.43	21.22	**18.98**	19.67
Motorcycle	9.87	9.58	9.01	8.88	8.66	9.02	**8.45**
MotorcycleE	9.72	9.27	8.93	8.60	7.78	**7.69**	8.47
Piano	16.20	14.65	14.01	**13.53**	17.55	14.88	14.02
PianoL	33.44	33.10	30.25	30.61	31.41	**26.65**	29.32
Pipes	10.45	10.87	9.96	10.49	12.38	11.67	**9.85**
Playroom	22.68	19.41	16.62	15.63	23.98	16.25	**15.18**
Playtable	41.89	20.04	39.29	20.77	36.67	18.34	**16.71**
PlaytableP	13.81	12.10	13.08	11.25	19.91	**10.55**	10.98
Recycle	10.52	10.75	8.38	9.16	11.44	9.18	**8.28**
Shelves	39.52	34.52	36.11	31.92	41.13	37.88	**31.63**
Teddy	7.18	6.60	6.33	**5.27**	8.24	6.26	5.57
Vintage	33.22	29.73	30.05	29.83	33.53	27.30	**26.97**
Weighted Average	16.47	14.82	14.66	13.53	17.06	13.69	**12.94**

Table 2. Comparison of the weighted average error rates in the all-region between seven algorithms (%). All of the results are obtained without the refinement procedure except SRDD [13]. The lowest error records are marked in bold.

Image Sets	FCVF	CS-FCVF	PGIF	CS-PGIF	DSG	SRDD	Proposed
Adirondack	10.09	9.64	8.39	8.82	15.58	8.77	**7.96**
ArtL	21.90	22.02	21.64	21.23	30.79	22.86	**21.17**
Jadeplant	34.51	35.87	33.45	33.89	38.02	**32.47**	33.00
Motorcycle	13.50	13.34	13.23	13.18	17.52	13.54	**12.62**
MotorcycleE	13.68	13.13	13.25	13.17	16.68	**11.96**	12.92
Piano	19.99	18.61	18.29	**17.74**	23.48	19.30	18.11
PianoL	36.43	36.14	33.91	34.13	36.15	**30.49**	33.09
Pipes	**21.45**	22.01	21.52	22.10	26.28	22.88	21.49
Playroom	30.61	27.88	25.38	24.32	33.93	25.11	**23.79**
Playtable	44.65	24.39	42.63	25.57	42.70	23.29	**21.23**
PlaytableP	17.29	15.00	17.55	15.43	27.74	14.39	**14.28**
Recycle	12.30	12.58	10.50	11.59	17.50	11.51	**10.33**
Shelves	40.14	35.43	36.89	32.91	44.43	39.15	**32.64**
Teddy	12.59	12.14	11.94	**10.96**	17.48	11.75	11.19
Vintage	37.18	33.86	34.01	33.67	38.47	31.51	**30.94**
Weighted Average	21.74	20.26	20.49	19.47	26.31	19.54	**18.71**

Table 3. Comparison of the computation time for four selected image sets between five algorithms (s).

Methods	Adirondack	Playroom	Playtable	Shelves
FCVF	15	16	14	15
CS-FCVF	20	21	18	18
PGIF	23	24	22	22
CS-PGIF	28	29	26	26
Proposed	31	31	29	29

4. Conclusions

In this work, we propose a novel stereo matching scheme to make use of hierarchical pattern information in stereo matching. The scheme exploits feature with different level of scales for matching metrics.

Inspired by the scheme of Zhang [29] for multi-scale disparity cost aggregation, the scheme uses a hierarchy of parameters of the GIF-based linear models and exploits the pervasive guided-image-filtering [25] for efficient matching cost calculation. The resultant multi-scale features are collected to form an improved cost volume for disparity estimation by using a linear combination of the guidance image.

A performance evaluation of version 3 of the Middlebury stereo evaluation data set [31] showed that the proposed solution provided superior disparity accuracy and comparable processing speed when compared with the representative stereo matching algorithms. Besides, the scheme outperformed most of the state-of-art algorithms even without the refinement procedure.

Author Contributions: Conceptualization, C.Z. and Y.-Z.C.; Formal analysis, C.Z. and Y.-Z.C.; Funding acquisition, Y.-Z.C.; Investigation, Y.-Z.C.; Methodology, C.Z.; Project administration, Y.-Z.C.; Supervision, Y.-Z.C.; Writing—original draft, C.Z. and Y.-Z.C.; Writing—review & editing, Y.-Z.C.

Funding: This research was funded by the National Natural Science Foundation of China, under Grant No. 61471263; the Natural Science Foundation of Tianjin, China, under Grant No. 16JCZDJC31100; the Ministry of Science and Technology, Taiwan, R.O.C., under Grant No. MOST 107-2221-E-182-078 and MOST 108-2221-E-182-061; and Chang Gung Memorial Hospital, Taiwan, under Grant No. CORPD2H0011 and CORPD2J0041.

Conflicts of Interest: The authors declare no conflict of interest.

References

1. Scharstein, D.; Szeliski, R. A taxonomy and evaluation of dense two-frame stereo correspondence algorithms. *Int. J. Comput. Vis.* **2002**, *47*, 7–42. [CrossRef]
2. Zbontar, J.; LeCun, Y. Stereo matching by training a convolutional neural network to compare image patches. *J. Mach. Learn. Res.* **2016**, *17*, 1–32.
3. Nahar, S.; Joshi, M.V. A learned sparseness and IGMRF-based regularization framework for dense disparity estimation using unsupervised feature learning. *IPSJ Trans. Comput. Vis. Appl.* **2017**, *9*, 1–15. [CrossRef]
4. Mayer, N.; Ilg, E.; Hausser, P.; Fischer, P.; Cremers, D.; Dosovitskiy, A.; Brox, T. A large dataset to train convolutional networks for disparity, optical flow, and scene flow estimation. In Proceedings of the IEEE Conference on Computer Vision and Pattern Recognition, Las Vegas, NV, USA, 27–30 June 2016; pp. 4040–4048.
5. Pang, J.; Sun, W.; Ren, J.S.; Yang, C.; Yan, Q. Cascade residual learning: A two-stage convolutional neural network for stereo matching. In Proceedings of the IEEE International Conference on Computer Vision Workshops, Venice, Italy, 22–29 October 2017; pp. 878–886.
6. Kendall, A.; Martirosyan, H.; Dasgupta, S.; Henry, P.; Kennedy, R.; Bachrach, A.; Bry, A. End-to-end learning of geometry and context for deep stereo regression. In Proceedings of the IEEE International Conference on Computer Vision, Venice, Italy, 22–29 October 2017; pp. 66–75.
7. Chang, J.R.; Chen, Y.S. Pyramid stereo matching network. In Proceedings of the IEEE Conference on Computer Vision and Pattern Recognition, Salt Lake City, UT, USA, 18–22 June 2018; pp. 5410–5418.
8. Park, I.K. Deep self-guided cost aggregation for stereo matching. *Pattern Recognit. Lett.* **2018**, *112*, 168–175.
9. Pan, C.; Liu, Y.; Huang, D. Novel belief propagation algorithm for stereo matching with a robust cost computation. *IEEE Access* **2019**, *7*, 29699–29708. [CrossRef]
10. Yao, P.; Zhang, H.; Xue, Y.; Chen, S. As-global-as-possible stereo matching with adaptive smoothness prior. *IET Image Process.* **2019**, *13*, 98–107. [CrossRef]
11. Gupta, R.K.; Cho, S.Y. Stereo correspondence using efficient hierarchical belief propagation. *Neural Comput. Appl.* **2012**, *21*, 1585–1592. [CrossRef]
12. Hu, T.; Qi, B.; Wu, T.; Xu, X.; He, H. Stereo matching using weighted dynamic programming on a single-direction four-connected tree. *Comput. Vis. Image Underst.* **2012**, *116*, 908–921. [CrossRef]
13. Yin, J.; Zhu, H.; Yuan, D.; Xue, T. Sparse representation over discriminative dictionary for stereo matching. *Pattern Recognit.* **2017**, *71*, 278–289. [CrossRef]
14. Yoon, K.J.; Kweon, I.S. Adaptive support-weight approach for correspondence search. *IEEE Trans. Pattern Anal. Mach. Intell.* **2006**, *28*, 650–656. [CrossRef] [PubMed]
15. Yang, Q. Stereo matching using tree filtering. *IEEE Trans. Pattern Anal. Mach. Intell.* **2015**, *37*, 834–846. [CrossRef]
16. Pham, C.C.; Jeon, J.W. Domain transformation-based efficient cost aggregation for local stereo matching. *IEEE Trans. Circuits Syst. Video Technol.* **2013**, *23*, 1119–1130. [CrossRef]

17. Çığla, C.; Alatan, A.A. An efficient recursive edge-aware filter. *Signal Process. Image Commun.* **2014**, *29*, 998–1014. [CrossRef]
18. Yang, Q.; Li, D.; Wang, L.; Zhang, M. Full-image guided filtering for fast stereo matching. *IEEE Signal Process. Lett.* **2013**, *20*, 237–240. [CrossRef]
19. Hosni, A.; Rhemann, C.; Bleyer, M.; Rother, C.; Gelautz, M. Fast cost-volume filtering for visual correspondence and beyond. *IEEE Trans. Pattern Anal. Mach. Intell.* **2013**, *35*, 504–511. [CrossRef]
20. He, K.; Sun, J.; Tang, X. Guided image filtering. *IEEE Trans. Pattern Anal. Mach. Intell.* **2013**, *35*, 1397–1409. [CrossRef] [PubMed]
21. Li, Z.; Zheng, J.; Zhu, Z.; Yao, W.; Wu, S. Weighted guided image filtering. *IEEE Trans. Image Process.* **2014**, *24*, 120–129.
22. Kou, F.; Chen, W.; Wen, C.; Li, Z. Gradient domain guided image filtering. *IEEE Trans. Image Process.* **2015**, *24*, 4528–4539. [CrossRef] [PubMed]
23. Hong, G.S.; Kim, B.G. A local stereo matching algorithm based on weighted guided-Image-filtering for improving the generation of depth range images. *Displays* **2017**, *49*, 80–87. [CrossRef]
24. Zhu, S.; Yan, L. Local stereo matching algorithm with efficient matching cost and adaptive guided image filter. *Vis. Comput.* **2017**, *33*, 1087–1102. [CrossRef]
25. Zhu, C.; Chang, Y.Z. Efficient stereo matching based on pervasive guided image filtering. *Math. Probl. Eng.* **2019**, *2019*, 1–11. [CrossRef]
26. Hu, W.; Zhang, K.; Sun, L.; Yang, S. Comparisons reducing for local stereo matching using hierarchical structure. In Proceedings of the IEEE International Conference on Multimedia and Expo, San Jose, CA, USA, 15–19 July 2013; pp. 1–6.
27. Tang, L.; Garvin, M.K.; Lee, K.; Alward, W.L.M.; Kwon, Y.H.; Abràmoff, M.D. Robust multiscale stereo matching from fundus images with radiometric differences. *IEEE Trans. Pattern Anal. Mach. Intell.* **2011**, *33*, 2245–2258. [CrossRef] [PubMed]
28. Li, R.; Ham, B.; Oh, C.; Sohn, K. Disparity search range estimation based on dense stereo matching. In Proceedings of the IEEE Conference on Industrial Electronics and Applications, Melbourne, Australia, 19–21 June 2013; pp. 753–759.
29. Zhang, K.; Fang, Y.; Min, D.; Sun, L.; Yang, S.; Yan, S.; Tian, Q. Cross-scale cost aggregation for stereo matching. In Proceedings of the IEEE Conference on Computer Vision and Pattern Recognition, Columbus, OH, USA, 23–28 June 2014; pp. 1590–1597.
30. Moritz, M.; Geiger, A. Object Scene Flow for Autonomous Vehicles. In Proceedings of the Conference on Computer Vision and Pattern Recognition (CVPR), Boston, MA, USA, 8–10 June 2015; pp. 1–10.
31. Scharstein, D.; Hirschmüller, H.; Kitajima, Y.; Krathwohl, G.; Nesic, N.; Wang, X.; Westling, P. High-resolution stereo datasets with subpixel-accurate ground truth. In Proceedings of the German Conference on Pattern Recognition (GCPR 2014), Münster, Germany, 12–15 September 2014; pp. 1–12.

applied sciences

MDPI

Article

Automatic Error Compensation for Free-Form Surfaces by Using On-Machine Measurement Data

Wei-chen Lee *, Yu-tzu Lee and Ching-Chih Wei

Department of Mechanical Engineering, National Taiwan University of Science and Technology, Taipei 10607, Taiwan

* Correspondence: wclee@mail.ntust.edu.tw; Tel.: +886-2-2737-6478

Received: 15 June 2019; Accepted: 28 July 2019; Published: 30 July 2019

Featured Application: The algorithm of the research can be applied to machining industry to reduce the variation of the product dimensions.

Abstract: Currently, most computer numerical control (CNC) controllers lack the function needed to compensate machining errors for free-form surfaces. The objective of this research was to enhance the accuracy and precision of the machined free-form surfaces of a workpiece using the mirror compensation method with the on-measurement data. By mirroring the points measured after semi-finishing, a new free-form surface for finishing machining can be automatically reconstructed. The surface can then be used to generate the cutting tool path to reduce the errors during finishing machining. In this research, three different types of surfaces were used for evaluating the proposed method. The results show that the proposed method reduced the standard deviations of the three surface geometries by 61%, 61%, and 32%, respectively. We also evaluated the tool radius modification method commonly used in the industry for error compensation and found that there is no substantial reduction on standard deviation. Therefore, the effectiveness of the error compensation method proposed in this research is evident.

Keywords: error compensation; free-form surface; on-machine measurement; mirror compensation method

1. Introduction

Reducing machining errors is a common goal for the machining industry. For a three-axis machine tool, there are 21 different geometric errors, including the linear positioning errors for each axis and the angular errors for each axis as well as between two axes [1]. The linear positioning errors are usually caused by the backlash of the ball screws used for the linear axes. The errors can be minimized by calibration with the use of laser interferometry. In addition to the geometric errors, the dynamic errors generated during the machining process also affect the accuracy of machining. The errors are usually caused by the variable cutting forces, machine vibrations, and the uncertain wear of the cutting tools [2]. Generally, the computer numerical control (CNC) controllers of a machine tool cannot cope with these errors. The heat generated during machining can also affect the machining accuracy [3].

In today's industry, the inaccuracy of the free-form surfaces of a machined workpiece can be improved by the following procedure: Perform roughing machining and semi-finishing machining first. For the semi-finishing machining, the excess material, or the stock, will be increased so that there is enough material left for the finishing machining. The average error can be obtained through the measurement, which can be on-machine or off-machine, after the semi-finishing machining. Based on the error, we can modify the tool radius in the computer-aided manufacturing (CAM) software and then regenerate the tool paths for finishing machining. By compensating for the error by modifying the tool radius, the tool paths can only be shifted inward or outward by a constant value, and it is not possible to improve the accuracy at a specific location only, as illustrated in Figure 1. As shown in Figure 1b, the profile of the workpiece after semi-finishing machining is different from the target shape shown in Figure 1a. Even if we perform the compensation by tool radius modification as described above, the average geometry may be close to the target geometry, as shown in Figure 1c. However, the local variation cannot be improved.

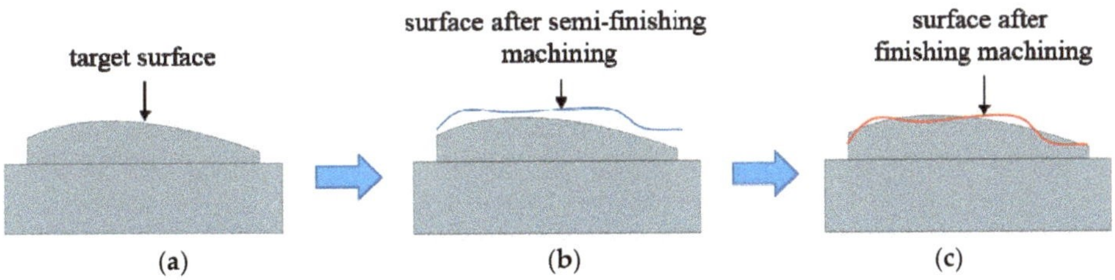

Figure 1. Illustration of the surface after semi-finishing machining and after finishing machining; (**a**) target surface of machining; (**b**) the profile after semi-finishing compared to the target surface; (**c**) the profile after finishing compared to the target surface.

Off-machine and on-machine measurement have advantages and disadvantages. To use off-machine measurement, it is not necessary to consider the measurement error caused by the machine tool itself; however, because the workpiece needs to be unmounted, there will be re-clamping errors. On the other hand, using on-machine measurement needs to rely on the repeatability of the machine tool, but it will not cause re-clamping errors, and can also reduce the time and efforts for re-clamping. At present, the precision of the machine tool is increasing, and the difference between the on-machine and the off-machine measurement is getting smaller. Therefore, dealing with the use of on-machine measurement results for error compensation has become an increasingly important issue.

Many error compensation methods have been proposed for free-form surfaces. Poniatowska [4] proposed a method to compensate for systematic errors on the machined surface. A machining pattern model (MPM), which represents the average of several surfaces machined under repeatable conditions, was used as the compensation basis. Regression analysis and spatial statistical methods were used to remove the random effects. The model was then used for reconstructing the computer aided design (CAD) model of the surface. The new cutting tool path can be generated to remove systematic errors. Chen et al. [5] used the spatial independence analysis method to decompose the machining errors measured online into systematic errors and random errors. For systematic errors, the numerical control (NC) codes can be modified for error compensation. Lai et al. [6] used the T-spline surface with local refinement instead of the common non-uniform rational B-splines (NURBS) surface to obtain better accuracy. The points used for the T-spline surface were obtained by using the on-machine measurement, and then the mirror symmetry model was adopted to construct the compensation surface. Cho et al. [7] performed machining surface error compensation through an on-machine measurement system. They performed multiple machining to obtain the error distribution. Based on the relationship between cutting conditions and surface error, an attempt was made to correct the tool path in order to reduce the errors using an iterative algorithm. Lasemi et al. [8] classified the errors in the machining of free-form surfaces as machine-related errors and process-related errors. The machine-related errors can be compensated offline, and the process-related errors can only be compensated online by re-planning the tool paths. Gu et al. [9] developed the global offset method to compensate for the machining errors of a five-axis machine tool. The method was based on the off-machine measurement of machined parts by using a coordinate measuring machine (CMM), and a compensation processor was used to calculate the parameters for the controller of the machine tool to perform compensation. Chen et al. [10] decomposed the machining errors into systematic errors and random errors by using the empirical mode decomposition method, and the systematic errors can then be compensated by using the offline measurement data. Huang et al. [11] developed an iterative compensation algorithm to modify the NC code based on the geometric errors, so that the real tool path generated by the compensated NC code and the designed tool path were within the tolerance. Zhong et al. [12] established the error of kinematics synthesis model and used the model to compensate the x-axis and z-axis positioning errors of a four-axis polishing machine. Msaddek et al. [13] presented a method of compensation by the insertion of the nodes to reduce the errors caused by the Bspline and Cspline interpolations on the machining surface. Chen [14] used an artificial neural network to predict the thermal error and used a PC-based controller to send the compensation information to the CNC controller to perform real-time error compensation.

Because the on-machine measurement depends on the precision of the machine tool, the source of the error on the machine tool is also critical. Biral et al. [15] used the online servo-controlled inclinometers to measure the geometric errors of the motion axes. That information was used by the CNC controller to compensate for the positioning errors. Guiassa et al. [16] used on-line measurement data for achieving compensation. Dimensional compensation can be achieved by correcting the tool size and the tool path. They proposed a method based on the cutting depth distribution to correct the tool offset error.

Regarding manufacturing the free-form surfaces, some novel methods have been proposed. Bo et al. [17] proposed an algorithm to find patches on free-form surfaces that can be machined by the flank of a conical cutting tool, and they extended their work to find the initial path for five-axis flank CNC machining with a general cutting tool [18]. Calleja et al. [19] also developed an algorithm to generate the flank milling paths for free-form surface machining with the use of a tapered mill tool. Using the flank of a cutting tool can increase the efficiency of manufacturing the free-form surfaces. Because gear surfaces may be treated as free-form surfaces, we can also find the information related to making free-form surface from gear manufacturing. Álvarez et al. [20] presented machining strategies and parameters used in different stages for milling the gear surfaces of spiral bevel gears by using NX CAD/CAM software.

From the previous research, we can learn that researchers have performed much compensation research based on the on-machine and off-machine measurement of free-form surfaces. However, determining how to reduce the time needed to generate the tool path for compensation and increase the accuracy are still in demand. In the past, the research often relied on CAIP (Computer-Aided Inspection Planning) software, but it was impossible to integrate machining, measurement, and compensation into the same software system, which would cause unnecessary difficulty for process automation. Therefore, the objective of this research was to integrate the machining, measurement, and compensation into the same CAD/CAM software to correct the machining error of free-form surfaces. It was concluded that the generation of the tool path for compensation could be very efficient, and the standard deviation of the machined free-form surfaces can be reduced by at least 32% using the mirror compensation method proposed in this paper.

2. Materials and Methods

2.1. Equipment and Software

The machine used in this research is a five-axis CNC vertical milling machine tool (model No. UX300) made by Quaser (Taiwan). The machine is table–table type, and its controller is SIEMENS 840D, which is connected to the Intranet in our institution. The on-machine measurement probe is the one with model number OMP400 made by RENISHAW (UK). The diameter of the stylus is 6 mm, and the length is 50 mm. The repeatability of the probe is 0.25 μm for two standard deviations.

The CAD/CAM system used in this study was NX (SIEMENS, Munich, Germany), and the program we developed for this research was written in Visual Studio C# (Microsoft, Redmond, WA, USA). We used the NX Open API to link NX with our program. As long as the measurement tool path is determined in NX, then the post-processor in NX can be used to output an NC file containing RENISHAW macro commands for probing. The NC file can then be combined with the NC file for manufacturing in the SIEMENS controller so that the on-machine measurement can be performed after the material cutting. After the on-machine measurement is completed, the measured data are stored in the controller, and our program can access the data through the Intranet for the subsequent processes.

2.2. Free-Form Surface Design

The workpiece material for the experiment was A6061-T6 aluminum alloy, and the size was 75 × 63 × 30 mm. In this study, three different surfaces were adopted for the experiment. For each surface, we machined four replicas so that we could obtain more measurement data. There are three profiles mainly used in the free-form surface design, as shown in Figures 2–4, all of which are quadratic curved surfaces. The first surface, Surface one, is shown in Figure 2. The curves in the u-direction on the curved surface are identical quadratic curves, and those in the v-direction are horizontal straight lines.

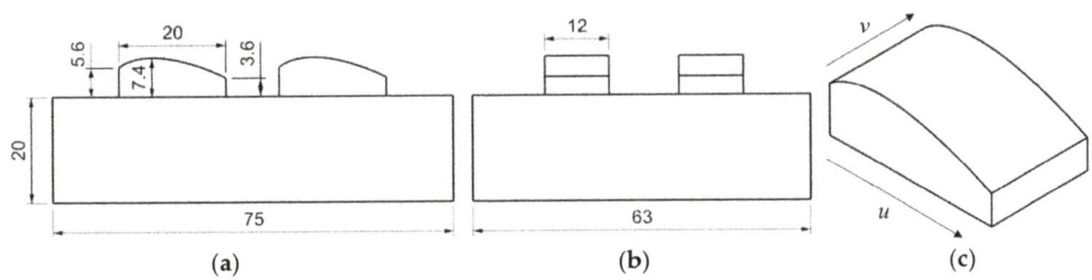

Figure 2. Design of Surface one; (**a**) front view; (**b**) side view; (**c**) isometric view.

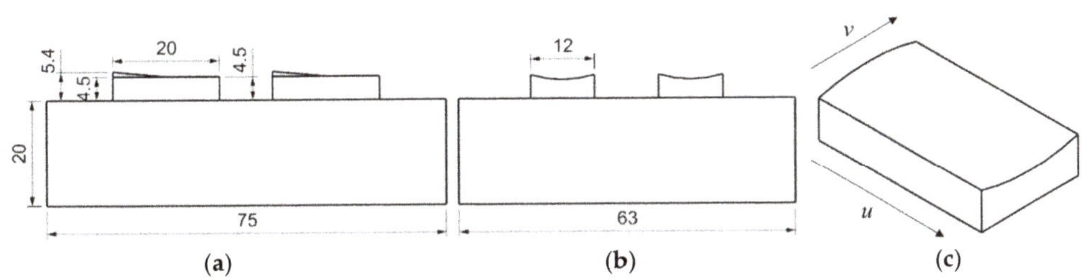

Figure 3. Design of Surface two; (**a**) front view; (**b**) side view; (**c**) isometric view.

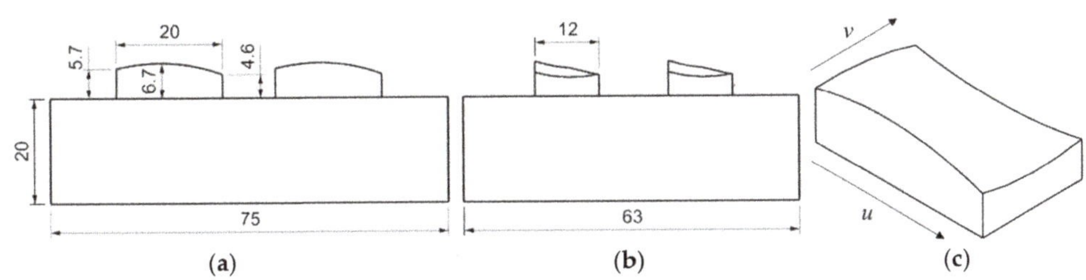

Figure 4. Design of Surface three; (**a**) front view; (**b**) side view; (**c**) isometric view.

The second surface, Surface two, is shown in Figure 3. The curves in the u-direction on the surface are straight lines with different slopes. The ones in the v-direction are quadratic curves, which start from a convex quadratic curve from one side and end with a concave curve at the other side.

The third surface, Surface three, shown in Figure 4, is the most complex surface among the three surfaces. The curves in the u-direction and the v-direction on the surface are all quadratic curves. They all start from convex quadratic curves from one side, and end with concave curves at the other side.

2.3. Measurement Points and Fitted Free-Form Surfaces

The measurement points for generating the measurement paths were all set on the ideal surface in the original CAD model. The number of measurement points is related to the accuracy of the surface fitting. The locations of the measurement points were determined based on the slopes of the points on the curves along the u-direction. We added measurement points at locations where the change of the slope of two adjacent measurement points was significant, and removed the measurement points at locations where the change of the slope of two adjacent measurement points was little. Figure 5 shows the measurement points on Surface one. Seven curves (not shown) in the u-direction were used, and ten measurement points are distributed along each curve. From this figure, we can see that there were more measurement points on the top-left of the surface because that was the place where the slopes of the points on the curves changed rapidly.

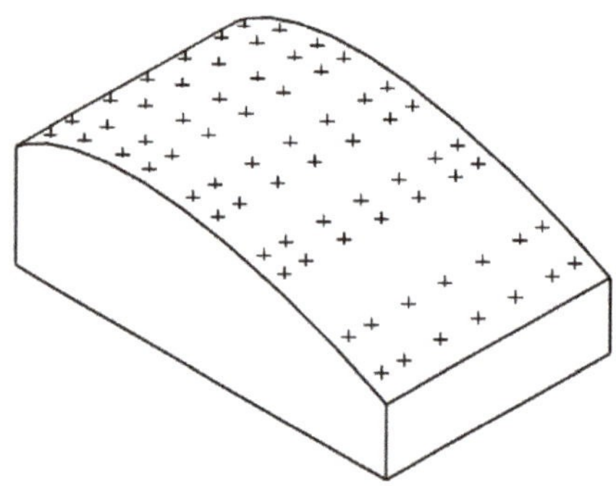

Figure 5. Illustration of the measurement points on Surface one.

In this study, we used the functions in NX Open API to perform both the curve fitting and the surface fitting. After obtaining the compensation points, we used the points on the same v coordinate to perform curve fitting along the u-direction, and the points on the same u coordinate to perform curve fitting along the v-direction. The two groups of curves were then used for surface fitting. The order of the fitted curve was determined using the criteria that the average of the distances from the compensation points to the fitted curve must be less than a threshold, which was set at 1 μm in the study. It was found that we could use the third-degree curves to achieve this.

NX includes at least five surface fitting methods: Studio Surface, Fit Surface, Through Curves, Through Curve Mesh, and UFun. To evaluate the methods, we designed two experiments: One was to use the five methods with a set of the compensation points for each surface design to fit surfaces, and find the average of the distances, d_1, from the compensation points to the fitted surface. The other was to arbitrarily pick up 1000 points from the surfaces of the reconstructed CAD models, and find the average of the distance, d_2, of these points to the fitted surface used for the reconstruction of the CAD models. The experimental results are listed in Table 1. It can be seen that the Through Curves Mesh method had the best results for Surface two and Surface three. For Surface one, the Through Curve Mesh performed second to UFun. However, UFun did not perform well for Surface two and Surface three. Therefore, in the subsequent study, we used Through Curve Mesh for surface fitting.

Table 1. Comparison of the five surface fitting functions in NX. Unit: μm.

	Studio Surface	Fit Surface	Through Curves	Through Curve Mesh	UFun
Surface One					
d_1	0.3	0.8	0.3	0.3	0
d_2	0.4	4.1	0.4	0.4	0.4
Surface Two					
d_1	0	0.1	0	0	0
d_2	0.5	0.6	0.8	0.2	0.8
Surface Three					
d_1	0	0.2	0	0	0
d_2	0.4	3.0	0.6	0.1	0.6

2.4. *Concept of Mirror Compensation*

In this study, the mirror compensation method was used to compensate for the errors of free-form surface machining. The compensation method was used after the semi-finishing machining was completed. The on-machine measurement was made on the semi-finishing surfaces to obtain the actual semi-finishing surface. Then, a mirror surface was introduced to mirror the actual semi-finishing surface to obtain the finishing surface for error compensation. The mirror surface is the mid-surface between the ideal semi-finishing surface and the ideal finishing surface, as indicated in Figure 6. This method finds where there is an undercut and where there is an overcut after semi-finishing machining. As illustrated in Figure 6, an undercut indicates that the remaining amount of material has exceeded the expected amount of material; an overcut indicates that the remaining amount of material has fallen short of the expected amount of material. Afterward, the semi-finishing surface is mirrored. The undercut locations on the semi-finishing surface will receive overcut during the finishing machining; the overcut locations will receive undercut during the finishing machining. As a result, after finishing machining, the machined surface is expected to approach the expected ideal surface or the target surface.

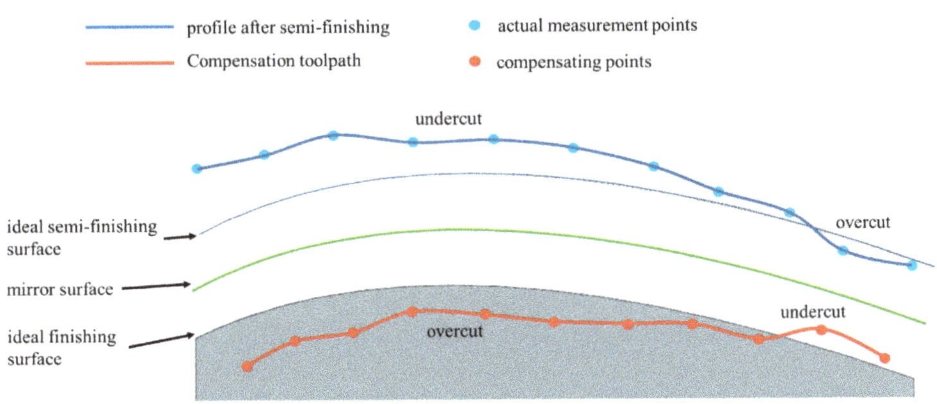

Figure 6. Illustration of the mirror compensation method.

How to use the mirror compensation method to find the compensation points is illustrated in Figure 7. First, we defined our ideal point P_P on the ideal finishing surface S_F. Then, we could obtain the normal vector $\mathbf{V_N}$ at P_P on the ideal finishing surface S_F. As explained previously, the mirror surface is the mid-surface between the ideal semi-finishing surface and the ideal finishing surface. Therefore, the distance D_H, which is half of the stock for the semi-machining, from P_P to the mirror surface is known. Then Equation (1) can be used to find P_M on the mirror surface. As long as P_M is known, the distance D_M from P_M to P_R can be readily obtained, as can the compensation point P_C by using Equation (2).

$$P_M = P_P + D_H \mathbf{V_N}. \tag{1}$$

$$P_C = P_R - 2D_M \mathbf{V_N}. \tag{2}$$

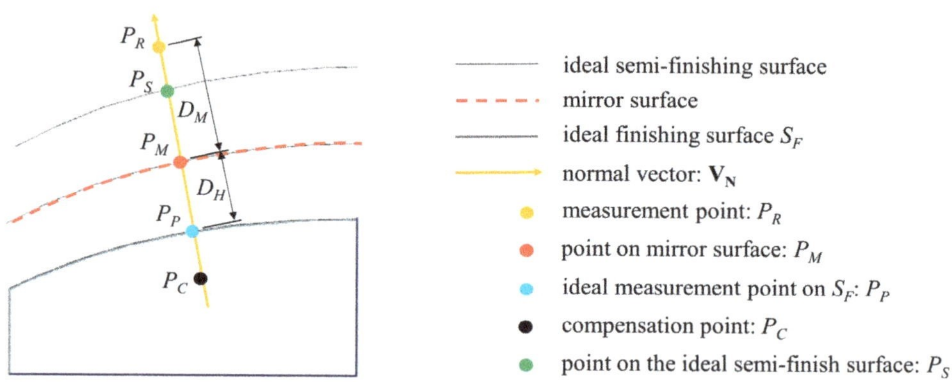

Figure 7. The locations of the points used for the mirror compensation method.

In the following analysis, the distance between P_R to P_P was used as the error for semi-finishing machining, and the distance from the actual measurement point after finishing machining to point P_P was used as the error for finishing machining.

2.5. Automatic Compensation Implementation

The complete compensation process is shown in Figure 8. We used the C# program we developed to import the on-machine measurement results. Because we used the same work coordinate system created by G54 command for cutting tool path and the on-machine measurement, the on-machine measurement data can be registered with the CAD model used for generating the G code for machining. Then, the mirror compensation method can be applied to find the locations of the compensation points. It follows that the curve fitting is performed on the compensation points, and the free-form surface can then be constructed by using the Through Curve Mesh fitting function in NX. Finally, the fitted surfaces will replace the original surfaces in the CAD model so that the model is reconstructed for generating the finishing tool path. The program we developed can automatically execute the entire process. For a model like the one shown in Figure 8, it took only about 30 s to complete the processes on a regular PC. It may take a much longer time to complete the same task by manual operation.

The complete machining process, including error compensation, is shown in Figure 9. First, the roughing machining and semi-finishing machining are performed. Then the on-machine measurement is performed, and the measurement data are recorded inside the CNC controller. Next, the automatic compensation program is used to perform the procedure, as illustrated in Figure 8. After obtaining the finishing tool path, we can perform the finishing machining to complete the machining with error compensation.

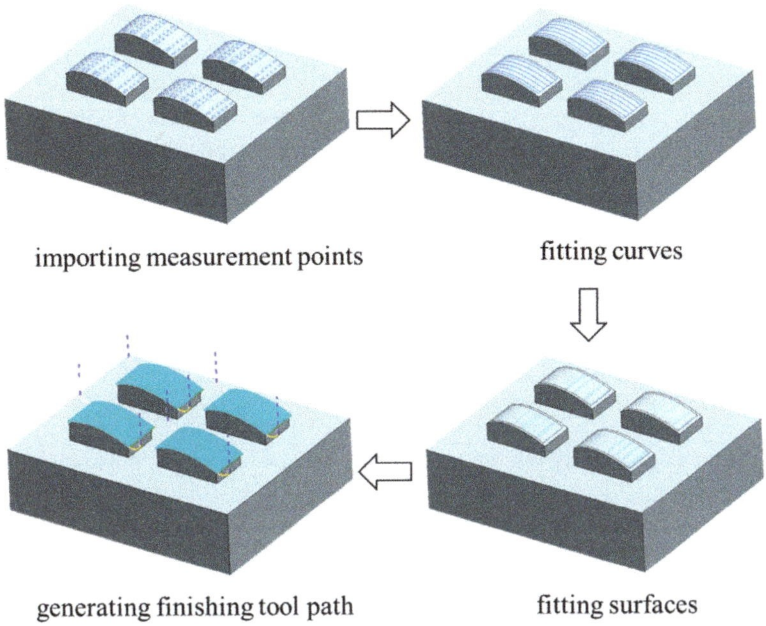

Figure 8. Automatic compensation flow for free-form surfaces.

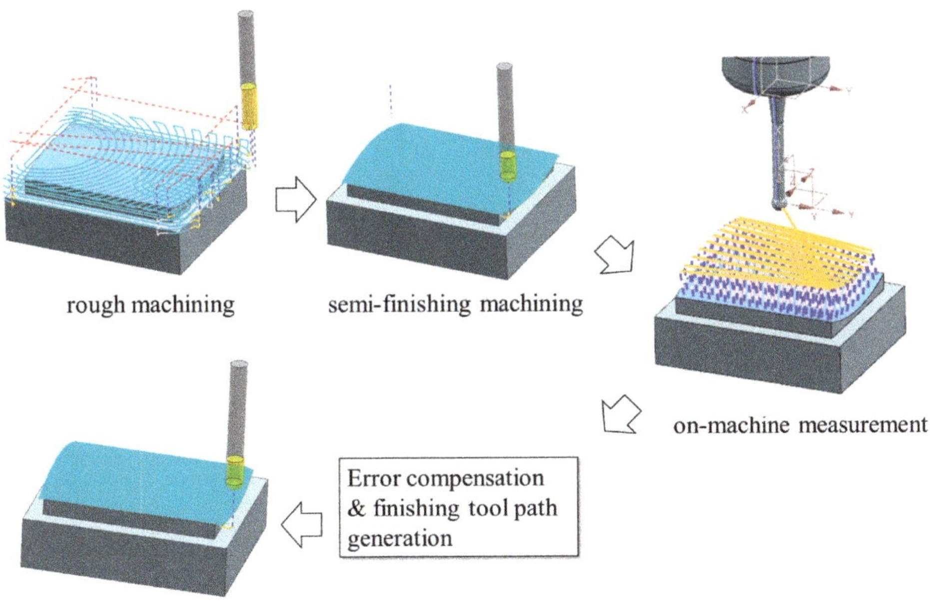

Figure 9. Complete machining processes with error compensation.

2.6. CAM Setup and Cutting Conditions

The tool paths for machining the free-form surfaces were generated in NX. The stock for the rough machining, the semi-finishing machining, and the finishing machining was set at 0.4 mm, 0.2 mm, and 0 mm, respectively. For the semi-finishing machining and finishing machining, the streamline operation in NX was adopted for tool path generation. Also, the scallop height was set at 1 μm, and the profile error, which is the chordal error along the linear interpolated tool path, was set at ±1 μm.

The cutting tool used in free-form surface machining for both the semi-finishing and finishing machining was made of tungsten carbide with TB coated. The tool was a two-flute ball end mill, 6 mm in diameter. The cutting conditions were as follows: the spindle speed was 10,000 RPM, the feed rate was 1200 mm/min, the surface cutting speed was 188 m/min, and the corresponding feed rate per blade was 0.06 mm/tooth.

3. Results and Discussion

This section presents the results of the automatic error compensation method proposed in this study. The number of measurement points of each replica for the three surface designs was different. Surfaces one, two, and three had 70, 49, and 68 measurement points for each replica, respectively. We used two different error compensation methods, tool radius modification and the mirror method as discussed previously, to carry out the error compensation and find their results.

Because we machined four replicas for each surface design, as shown in Figures 2–4, we would like to examine the repeatability of the four replicas first. Table 2 shows the standard deviations of the errors of the four replicas for each surface design. The values without the parentheses represent the standard deviations for the replicas before and after error compensation, and the values within the parentheses represent the range (the difference between the lowest and highest values) for the standard deviation of the errors for the four replicas. As listed in Table 2, most of the ranges are less than 0.5 μm, and the maximum range is 0.8 μm, which comes from the measurement results of Surface three after mirror compensation. Except for this case, all the other results show consistent standard deviations among the four replicas for the three surfaces before and after compensation, regardless which compensation method we used.

Table 2. Standard deviations of the errors of the four replicas for the three surfaces by using tool radius compensation or mirror compensation. Each of the values within the parentheses represents the range of the four values above it. Unit: μm.

	Tool Radius Compensation		Mirror Compensation	
Surface Design	**Before**	**After**	**Before**	**After**
Surface one	6.0, 6.1, 6.3, 6.1 (0.3)	6.2, 6.4, 5.8, 6.2 (0.6)	6.0, 6.1, 6.3, 6.3 (0.3)	2.0, 1.8, 2.0, 1.9 (0.2)
Surface two	3.8, 3.7, 3.3, 3.8 (0.5)	3.4, 3.4, 3.7, 3.3 (0.4)	3.8, 3.4, 3.8, 3.9 (0.5)	1.1, 1.5, 1.3, 1.2 (0.4)
Surface three	2.9, 2.9, 3.0, 3.0 (0.1)	3.2, 3.2, 3.2, 3.0 (0.2)	3.0, 3.0, 2.9, 3.0 (0.1)	1.4, 2.2, 2.0, 2.0 (0.8)

Because the repeatability of the four replicas is reasonably good in general, we will show the error distribution of one of the replicas (replica one) before and after error compensation by using either the tool radius compensation method or the mirror compensation method to illustrate the error distribution for all four replicas. The reason we do not show the distribution for all four replicas is that we would like to focus on the comparison between the two compensation methods without using too much space of the paper.

After performing the semi-finishing machining on the four replicas of Surface one, we obtained the error distribution of the free-form surface of replica one, as the circles shown in Figure 10a. The average error after semi-finishing machining was 12.3 μm with a standard deviation of 6.0 μm. After the tool radius compensation was performed, the error distribution became the plus signs in Figure 10a. The average error was reduced to 3.1 μm, but the standard deviation became 6.2 μm, which was almost unchanged. It is evident that the average error was much improved, but with no significant improvement in the variation.

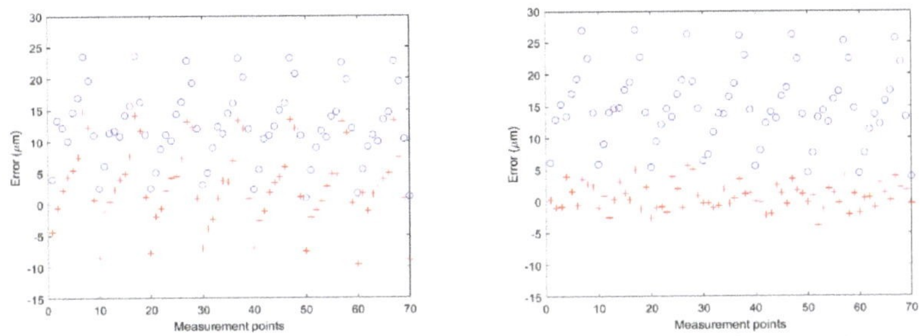

(a) (b)

Figure 10. (**a**) The error distribution of replica one before (blue circles) and after (red plus signs) tool radius compensation for Surface one; (**b**) the error distribution of replica one before (blue circles) and after (red plus signs) mirror compensation for Surface one.

On the other hand, we repeated the previous experiment by using the mirror compensation method proposed in this study. The error distribution of replica one before and after the mirror compensation is shown in Figure 10b. Because the error compensation was already performed on the free-form surfaces used in the previous experiment, we had to machine new surfaces for this experiment. The average error after semi-finishing machining was 15.0 μm with a standard deviation of 6.0 μm. After error compensation was performed, the average error was 0.7 μm, and the standard deviation was 2.0 μm. The average error was improved by 14.3 μm, and the standard deviation was improved by 4.0 μm, which demonstrated that the proposed error compensation method could significantly improve the surface variation. Similar results were obtained for the other three replicas of Surface one.

Similar to the experiments on Surface one, we also experimented with Surface two and Surface three. Figures 11 and 12 show the error distributions for one of the replicas before and after compensation on Surfaces two and three, respectively.

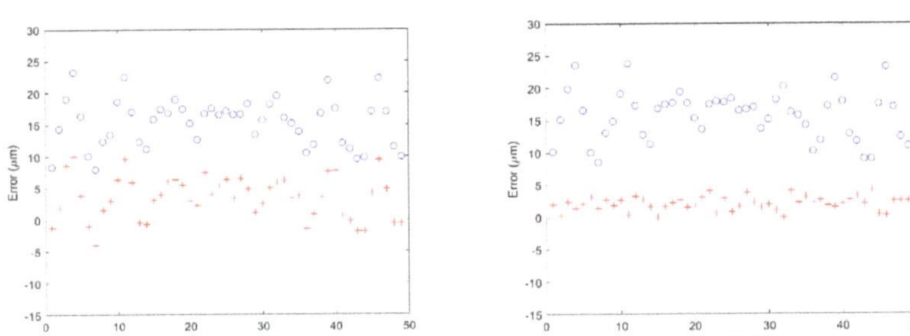

(a) (b)

Figure 11. (**a**) The error distribution of replica one before (blue circles) and after (red plus signs) tool radius compensation for Surface two; (**b**) the error distribution of replica one before (blue circles) and after (red plus signs) mirror compensation for Surface two.

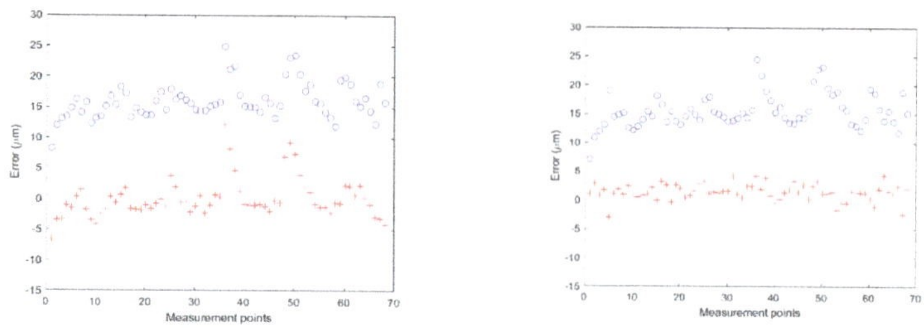

(a) (b)

Figure 12. (**a**) The error distribution of replica one before (blue circles) and after (red plus signs) tool radius compensation for Surface three; (**b**) the error distribution of replica one before (blue circles) and after (red plus signs) mirror compensation for Surface three.

The experimental results are summarized in Table 3. As mentioned previously, Surfaces one, two, and three had 70, 49, and 68 measurement points for each replica, respectively, and each surface design had four replicas, so the number of errors for Surface one, two, and three was 280, 196, and 272, respectively. We used the errors to calculate the mean and standard deviation of the errors for each surface design, and the results are listed in Table 3. For all the three different surfaces, it can be seen that both compensation methods can greatly improve the mean errors. The mean errors can be reduced to be within 2.0 μm or less for both methods. Regarding the standard deviations for Surfaces one, two, and three, the standard deviations after using mirror compensation improved by 3.8 μm, 2.3 μm, and 1.0 μm, which was equivalent to 61%, 61%, and 32%. The standard deviation can only improve by 0.1 μm by using the tool radius compensation for all three surfaces. It is evident that to improve the standard deviation of the error, the mirror compensation method proposed in the paper is more effective compared to the tool radius compensation method.

Table 3. Mean (the values without the parentheses) and the standard deviation (the values within the parentheses) of the errors for the three surfaces by using tool radius compensation or the mirror compensation, unit: μm.

	Tool Radius Compensation		**Mirror Compensation**	
Surface Design	**Before**	**After**	**Before**	**After**
Surface one	14.1 (6.2)	2.0 (6.1)	15.1 (6.2)	0.1 (2.4)
Surface two	17.2 (3.8)	0.9 (3.7)	17.2 (3.8)	0.7 (1.5)
Surface three	16.5 (3.2)	−0.3 (3.1)	16.8 (3.1)	0.3 (2.1)

Regarding the different types of surface, Surface one is the least complicated surface, and Surface three is the most complicated one, as discussed previously. Intuitively, we expected Surface three to have the largest standard deviation of the errors and Surface one the smallest before error compensation. However, the actual standard deviations of errors were not as expected. Surface one had the largest variation. The reason for this phenomenon is unknown.

We also used the surface profile per ISO standard 1101 as a measure for evaluating the two compensation methods, and the results are summarized in Table 4. For Surfaces one, two, and three, the surface profile error after using tool radius compensation could be improved by 24.8 μm, 32.3 μm, and 30.5 μm, which were equivalent to 45%, 62%, and 56%, respectively. However, using mirror compensation could improve the surface profile error by 44.8 μm, 44.4 μm, and 41.2 μm, which were equivalent to 78%, 84%, and 77%, respectively. For Surfaces one, two, and three, the mirror compensation method outperformed the tool radius compensation method by 33%, 22%, and 21%, respectively. It is obvious that for improving the surface profile error, the mirror compensation method proposed in the paper is also more effective than the tool radius compensation method.

Table 4. Surface profile error for the three surfaces by using the tool radius compensation or the mirror compensation, unit: μm.

	Tool radius Compensation		Mirror Compensation	
Surface Design	**Before**	**After**	**Before**	**After**
Surface one	54.9	30.1	57.6	12.8
Surface two	52.3	20.0	52.7	8.3
Surface three	54.8	24.5	53.2	12.0

4. Conclusions

An efficient and effective method to compensate for the errors of free-form surfaces is needed in the current machining industry. In this study, we developed a program based on mirror compensation by using the API function of NX, a CAD/CAM system, to integrate the machining, measurement, and compensation. After performing the semi-finishing machining, we used on-machine measurement to obtain the locations of the measurement points, which were then mirrored to generate the locations of the compensation points. The points were used to construct a new surface in CAD to allow NX to generate the tool paths for finishing machining. Five functions in NX for surface fitting were evaluated, and the function, Through Curve Mesh, performed the best. Experiments on three different free-form surfaces were conducted, and the average standard deviations of the errors of the compensated surfaces can be reduced by 61%, 61%, and 32%, respectively, compared to those of the uncompensated surfaces. Although the popular tool radius compensation method can reduce the mean of the errors, it can only reduce the standard deviation of the errors by a little. In terms of the surface profile error defined in ISO 1011, the mirror compensation method proposed in the paper outperformed the tool radius compensation method by at least 21%. Based on the results, the mirror compensation method proposed in the paper can provide an effective way to reduce the variation and surface profile error of the machined free-form surface.

The program can complete the error compensation and generate the tool paths for finishing machining in about 30 s on a regular PC, which means that the processing time and technical difficulty can also be reduced significantly.

Author Contributions: Conceptualization, W.-c.L., Y.-t.L., and C.-C.W.; methodology, W.-c.L. and Y.-t.L.; software, Y.-t.L. and C.-C.W.; validation, Y.-t.L.; formal analysis, Y.-t.L.; investigation, Y.-t.L.; resources, W.-c.L.; data curation, Y.-t.L. and W.-c.L.; writing—original draft preparation, W.-c.L. and Y.-t.L.; writing—review and editing, W.-c.L.; visualization, W.-c.L. and Y.-t.L.; supervision, W.-c.L.; project administration, W.-c.L.; funding acquisition, W.-c.L.

Funding: This research was supported by the Ministry of Science and Technology, Taiwan, Republic of China, under Grant number MOST 107-2218-E-011-002.

Conflicts of Interest: The authors declare no conflict of interest.

References

1. Lee, E.-S.; Suh, S.-H.; Shon, J.-W. A comprehensive method for calibration of volumetric positioning accuracy of CNC-machines. *Int. J. Adv. Manuf. Technol.* **1998**, *14*, 43–49. [CrossRef]
2. López de Lacalle, L.N.; Lamikiz, A.; Muñoa, J.; Salgado, M.A.; Sánchez, J.A. Improving the high-speed finishing of forming tools for advanced high-strength steels (AHSS). *Int. J. Adv. Manuf. Technol.* **2006**, *29*, 49–63. [CrossRef]
3. Mekid, S.; Ogedengbe, T. A review of machine tool accuracy enhancement through error compensation in serial and parallel kinematic machines. *Int. J. Precis. Technol.* **2010**, *1*, 251–286. [CrossRef]
4. Poniatowska, M. Free-form surface machining error compensation applying 3D CAD machining pattern model. *Comput. Aided Des.* **2015**, *62*, 227–235. [CrossRef]
5. Chen, Y.; Gao, J.; Deng, H.; Zheng, D.; Chen, X.; Kelly, R. Spatial statistical analysis and compensation of machining errors for complex surfaces. *Precis. Eng.* **2013**, *37*, 203–212. [CrossRef]

6. Lai, J.; Fu, J.; Xia, C.; Lin, Z.; Fu, G.; Chen, Z. Error compensation of free-form surface with critical area based on T-spline surface reconstruction. *Int. J. Comp. Integ. Manuf.* **2017**, *30*, 782–791. [CrossRef]
7. Cho, M.W.; Seo, T.I.; Kwon, H.D. Integrated error compensation method using OMM system for profile milling operation. *J. Mater. Process Tech.* **2003**, *136*, 88–99. [CrossRef]
8. Lasemi, A.; Xue, D.; Gu, P. Tool path re-planning in free-form surface machining for compensation of process-related errors. *Int. J. Prod. Res.* **2014**, *52*, 5913–5931. [CrossRef]
9. Gu, J.; Agapiou, J.S.; Kurgin, S. Error compensation and accuracy improvements in 5-axis machine tools using the global offset method. *J. Manuf. Syst.* **2017**, *44*, 324–331. [CrossRef]
10. Chen, Y.; Tang, H.; Tang, Q.; Zhang, A.; Chen, D.; Li, K. Machining error decomposition and compensation of complicated surfaces by EMD method. *Measurement* **2018**, *116*, 341–349. [CrossRef]
11. Huang, N.; Jin, Y.; Bi, Q.; Wang, Y. Integrated post-processor for 5-axis machine tools with geometric errors compensation. *Int. J. Mach. Tool. Manu.* **2015**, *94*, 65–73. [CrossRef]
12. Zhong, Z.E.; Ji, Z.; Jun, J.S.; Gang, L. Comprehensive error modelling and error compensation for complex optical free-form surface polishing platform. *Int. J. Nanomanuf.* **2017**, *13*, 371–383. [CrossRef]
13. Msaddek, E.B.; Baili, M.; Bouaziz, Z.; Dessein, G. Compensation of machining errors of Bspline and Cspline. *Int. J. Adv. Manuf. Technol.* **2018**, *97*, 4055–4064. [CrossRef]
14. Chen, J.-S. Computer-aided accuracy enhancement for multi-axis CNC machine tool. *Int. J. Mach. Tool. Manu.* **1995**, *35*, 593–605. [CrossRef]
15. Biral, F.; Bosetti, P. On-line measurement and compensation of geometrical errors for cartesian numerical control machines. In Proceedings of the 9th IEEE International Workshop on Advanced Motion Control, Istanbul, Turkey, 27–29 March 2006; IEEE: Piscataway, NJ, USA, 2006; pp. 120–125.
16. Guiassa, R.; Mayer, J.R.R.; St-Jacques, P.; Engin, S. Calibration of the cutting process and compensation of the compliance error by using on-machine probing. *Int. J. Adv. Manuf. Technol.* **2015**, *78*, 1043–1051. [CrossRef]
17. Bo, P.; Bartoň, M.; Pottmann, H. Automatic fitting of conical envelopes to free-form surfaces for flank CNC machining. *Comput. Aided Des.* **2017**, *91*, 84–94. [CrossRef]
18. Bo, P.; Bartoň, M. On initialization of milling paths for 5-axis flank CNC machining of free-form surfaces with general milling tools. *Comput. Aided Geom. Des.* **2019**, *71*, 30–42. [CrossRef]
19. Calleja, A.; Bo, P.; González, H.; Bartoň, M.; López de Lacalle, L.N. Highly accurate 5-axis flank CNC machining with conical tools. *Int. J. Adv. Manuf. Technol.* **2018**, *97*, 1605–1615. [CrossRef]
20. Álvarez, Á.; Calleja, A.; Ortega, N.; De Lacalle, L.N.L. Five-Axis Milling of Large Spiral Bevel Gears: Toolpath Definition, Finishing, and Shape Errors. *Metals* **2018**, *8*, 353. [CrossRef]

applied sciences

MDPI

Article

3D and Boundary Effects on 2D Electrical Resistivity Tomography

Yin-Chun Hung [1,*], Chih-Ping Lin [2], Chin-Tan Lee [3] and Ko-Wei Weng [3]

[1] Department of Urban Planning and Landscape, National Quemoy University, Kinmen 892, Taiwan
[2] Department of Civil Engineering, National Chiao Tung University, Hsinchu 300, Taiwan
[3] Department of Electronic Engineering, National Quemoy University, Kinmen 892, Taiwan
* Correspondence: hij@nqu.edu.tw; Tel.: +886-82-313786

Received: 11 June 2019; Accepted: 20 July 2019; Published: 24 July 2019

Abstract: Electrical resistivity tomography (ERT) is one of the most widely used geophysical methods in geological, hydrogeological, and geo-environmental investigations. Although 3D ERT is now available, 2D ERT remains state-of-the-practice due to its simplicity in fieldwork and lower space requirements. 2D ERT assumes that the ground condition is perpendicular to the survey line and outside the survey line is homogeneous. This assumption can often be violated in conditions such as geologic strikes not perpendicular to the survey line and topographic changes or buried objects near the survey line. Possible errors or artifacts in the 2D resistivity tomogram arising from violating the 2D assumption are often overlooked. This study aimed to numerically investigate the boundary effects on 2D ERT under various simplified conditions. Potential factors including resistivity contrast, depth and size of buried objects, and electrode spacing were considered for the parametric studies. The results revealed that offline geologic features may project onto the 2D tomogram to some extent, depending on the aforementioned factors. The mechanism and implications of boundary effects can be drawn from these parametric studies.

Keywords: electrical resistivity tomography; boundary effect; 3D effect

1. Introduction

Electrical resistivity tomography (ERT) applied to geotechnical investigation has developed rapidly in recent years with the beginning of 1D exploration that is mainly used in groundwater and mining exploration. With the advancement of computer technology and forward and inverse computing skills, in the past ten years, 2D exploration has been widely used in the investigation and monitoring of geotechnical engineering, environmental engineering, and groundwater pollution. 3D ERT is gradually starting to be used, however, due to the limitations of the test environment, 3D is the trend of future development, so the application in current engineering is mainly based on 2D ERT [1–7].

Linearity 2D ERT assumes that the geological formation resistivity property is homogeneous in the vertical line direction. When the geological formation does not meet the condition where they homogenize in the vertical line direction, since the electric current is flowing in 3D, the 2D ERT result may be in error, so the distortion that this hypothesis may cause is often ignored in the interpretation of the test results. The 3D effect means that the geological structure outside the 2D resistivity profile will map to the 2D resistivity profile and cause such errors. The boundary effect means that when 2D ERT is measured on a straight line of finite length but the boundary of the survey line encounters terrain changes such as pools, valleys, or underground structures like concrete structures or metal pipelines, it may be mapped to the 2D electrical resistivity profile, causing exploration errors, thus affecting the test results and the accuracy of the interpretation.

The application of ERT in geotechnical investigation has been developed quite completely, and there exist fairly complete procedures from the arrangement of the test to the inverse computation analysis of the data, but there has not been a detailed assessment of the 3D effect and the boundary effect. 2D ERT assumes that the geological formation is a 2D semi-infinite space. However, in the real state of geological formations, the electric current flows in the 3D (X, Y, Z) direction, therefore, the object in the non-2D section has a certain degree of disturbance to the ground resistivity electric field, which causes some irregular resistivity and noise on the 2D profile [8]. Torleif Dahlin explored the application of 2D ERT in the environment and engineering in Sweden. In the in situ test, it was considered that there was a 3D effect, and they suggested avoiding this when setting the line [9]. Lin et al. explored the detection of dam leakage by 2D ERT and found that the 3D effect was generated by the structure around the line and projecting the feature of the structure onto the testing profile. Therefore, they suggested that the effects of the 3D effect should be noted when testing [10]. Many scholars often face the boundary effect caused by the boundary of the sandbox when conducting an indoor sandbox experiment [11–16]. Mei, Xing-Tai used a homogeneous thickness model made from dough to explore the indoor test method and criteria for 2D ERT. As the limited boundary on the three sides of the sand-box will affect the transfer of the electric current, this is inconsistent with the assumption that the theoretical electric current has boundless extended flows in the underground half-space, and will consequently cause the boundary effect. After discussion, it was found that the boundary of the sand-box will cause the resistivity to be in an unstable state until the boundary of the bilateral line is more than double spread, where the apparent resistivity will be stable [17].

It can be found from the above literature that the 3D effect and the boundary effect are the problems that most scholars may face when conducting electrical resistivity tomography. However, few scholars have explored the 3D effect and the boundary effect in depth. The purpose of this study was to explore the 3D effect and boundary effect on 2D ERT and to understand the possible effects of changes in the geological formation outside the line. In this paper, through the establishment of a numerical model, we simulated 3D models and boundary stratigraphic fluctuation including a model where the extremity line was medium and encountered pipelines aside from the line. We used the parametric variation of the resistivity ratio, electrode spacing, depth of media embedding, and medium size to explore the influence situation of the 3D effect and boundary effect on 2D ERT. Finally, based on the numerical modeling results, we propose suggestions or precautions for future testing.

2. ERT Background

ERT mainly uses two current poles (C1, C2) and two potential poles (P1, P2) to collect measurements, as shown in Figure 1. The low frequency alternating current accesses the geological formation through current poles. Under different geometrical positions, potential poles and current poles are separately used to measure the potential difference and current. Since the calculation involves the distribution of the conductive medium in the formation, there are a large number of potential difference solutions and boundary conditions that are necessary to solve through forward modeling, as shown in Equation (1). The resistivity calculated by forward modeling is called apparent resistivity, which is usually not the true resistivity of the underground electrical formations. The true resistivity can be obtained by calculating the apparent resistivity through a suitable inverse computation [18].

$$\varrho_a = \frac{V_{P1} - V_{p2}}{I} \times \frac{2\pi}{\frac{1}{r1} - \frac{1}{r2} - \frac{1}{r3} + \frac{1}{r4}} = K\frac{\Delta V}{I} \tag{1}$$

where K is the geometrical arrangement parameter of the electrode; ΔV is the measured potential difference; and I is the circulating current intensity.

Apparent resistivity means that when taking measurements in real geological formations, the measured resistivity may transform by changing the electrode spacing and position as the resistivity of the formation may be heterogeneous. For example, if we were to lay dozens of electrode rods equally on the Earth's surface in advance, then constantly change the spacing and strafing right to measure by the current poles and potential poles aforementioned, it can obtain a pseudo-section as shown in Figure 1. Finally, the resistivity distribution of the representative geological formation should be obtained through inverse analysis. Inverse analysis generally adopts an optimization method because the resistivity value calculated by inverse computation is close to that of the true resistivity value [19]. The software of inverse analysis makes an initial value guess for the inverse model through the measured pseudo-section, and after presetting the initial inverse model and minimizing the difference between the model's reacting value and the measured data value through multiple operations. The result of the inverse at present is the resistivity distribution of the real geological formation. This study adopted the EarthImager inverse software developed by AGI to perform numerical simulations and calculations [20].

The simulated testing method in this study adopted dipole–dipole electrode arrangement. The electrode sequence of the dipole–dipole array was C2, C1, P1, P2, respectively. The two current poles form a dipole, and the two potential poles form another dipole, where C1C2 = P1P2 = a and C1P1 = na. When the parameter n is gradually increased, the resistivity of the formation changing from shallow to deep can be obtained. The high sensitivity of this arrangement is concentrated between the paired current poles and the potential poles, so horizontal formation changes are more suitable, while the vertical direction changes poorly.

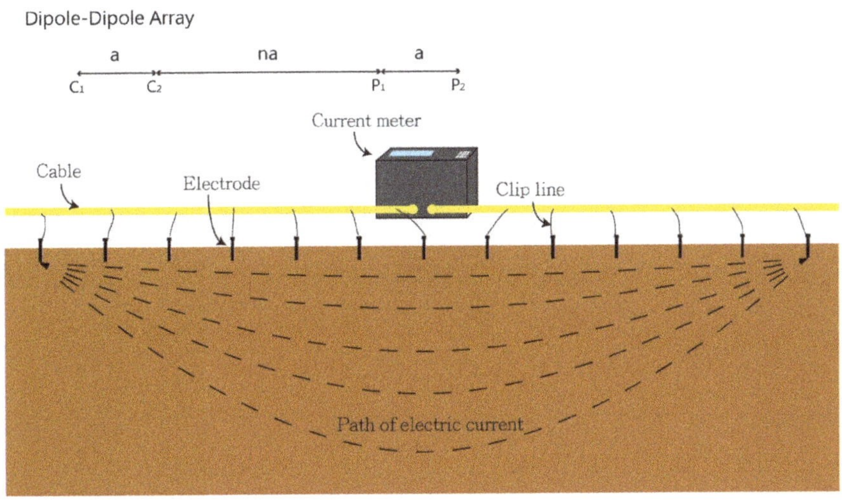

Figure 1. Schematic diagram of the ERT test.

3. Method

To understand the effects of the 3D effect and boundary effect on 2D resistivity profiles, this study separately established numerical models including models of pipelines buried next to the survey line and geological formation changes at the survey line boundary. Then, we investigated the possible effects on the 2D resistivity profile through the parametric variation such as the resistivity ratio, electrode spacing, depth of media buried, and medium size.

This study assumed that the resistivity ratio was $n = R2/R1$, where $R1$ is the resistivity in the background soil layer, and $R2$ is the soil layer outer resistivity profile or boundary medium. When $n > 1$, the boundary medium of high resistivity ($R2$) has little effect on the low resistivity zone ($R1$). When $n < 1$, the boundary medium of low resistivity ($R2$) has a large influence on the high resistivity zone ($R1$). The main reason is that the current is more concentrated and has an ability to flow better where the low resistivity zone is.

Therefore, this study will investigate the case when $n < 1$, that is, when the boundary medium is low resistivity, for the change of the high resistivity soil layer on the 2D resistivity profile.

3.1. 3D Effect Model with Pipeline Formation

To understand the boundary effect that may be caused by geological formations with pipelines, this study established a geological formation model with pipelines as shown in Figure 2. This study will establish and plan a geological formation model and testing parameters from the four aspects of resistivity ratio, spacing of the line and pipelines, depth of pipelines building, and pipeline size in the pipelines and strata.

In the model, the resistivity of the pipelines is *R2*, the soil layer resistivity is *R1*, *R1* > *R2*, and the ratio of the two is n ($n = R2/R1$). To discuss the influence of the resistivity ratio under this assumption, we fixed R1 as 1000 ohm-m, and set R2 as 10, 50, 100, and 250 ohm-m, respectively. The section size of the pipelines was $dyp \times dzp$, and the depth of the embedding was $dep = 2$ m (the top of the pipeline). The parallel pipelines around the pipeline setting were 5 to 6 lines, each line was separated $dy = 3$, the spacing to electrode rods was $dx = 3$, and each line had a slight adjustment to the length of line L with the sounding requirement. Table 1 presents the explanatory table of the 3D effect model parameter with the pipeline formation.

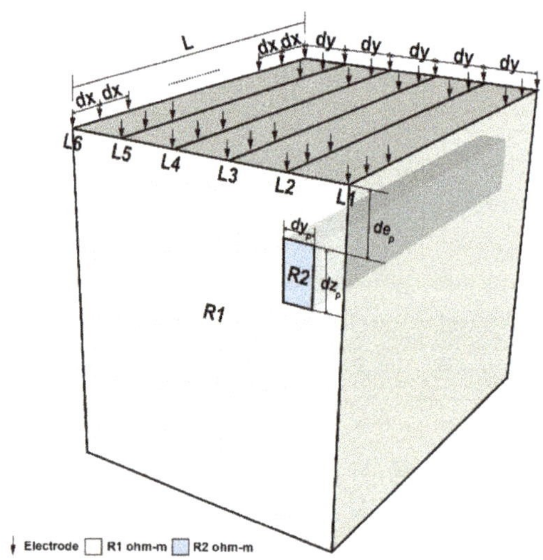

Figure 2. 3D effect model with the pipeline formation.

Table 1. The explanatory table of the 3D effect model parameters with the pipelines.

Model	Influence Factor	*R1* (ohm-m)	*R2* (ohm-m)	*dx* (m)	*dy* (m)	*dep* (m)	*dyp* (m)	*dzp* (m)	*L* (m)
3D effect model with pipeline	Resistivity ratio ($n = R2/R1$)	1000	10 50 100 250	3	3	2	1.5	2	42
	Pipeline size	1000	100	3	3	2	1.5 3	2 4	42 51
	Embedding depth	1000	100	3	3	0 4	1.5	2	42 51
	Influence distance	1000	10	3	3	2	1.5	2	42
	Electrode spacing	1000	10	1.5 3	3	2	1.5	2	42

3.2. Boundary Effect Model

To understand the boundary effects that may be caused by geological formation changes on both sides of the survey line, this study explored the effects of the boundary effect on 2D ERT through the changes in the boundary medium using numerical simulations by assuming there was heterogeneous media at the boundary. This study explored the influence difference through a model comparison with a boundary effect and no boundary effect, as shown in Figures 3 and 4.

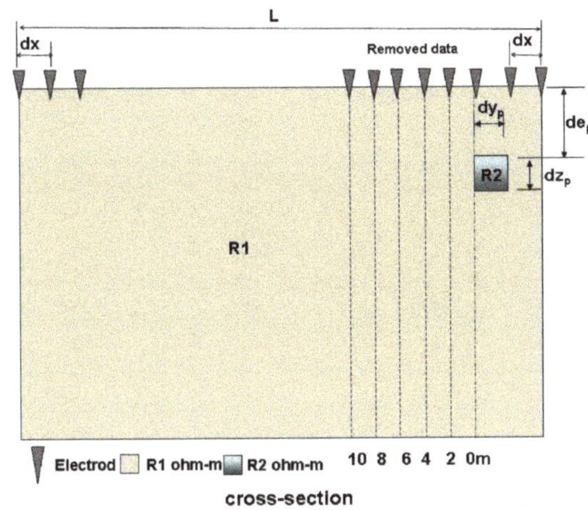

Figure 3. 2D boundary effect model.

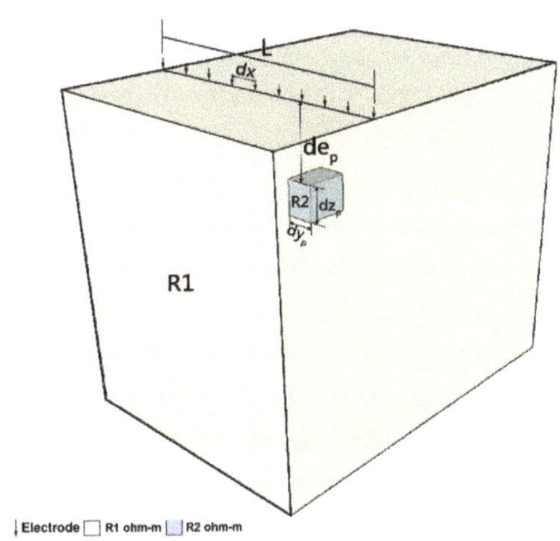

Figure 4. 2.5D boundary effect model.

The boundary effect was first calculated through the apparent resistivity profile of the model with a boundary medium by forward modeling. Then, we calculated the apparent resistivity by inverse calculation to obtain the resistivity profile of the model affected by the boundary medium. To understand the influence of the boundary effect, this study deleted the data points of the boundary medium in the apparent resistivity profile, and then through recalculation to obtain the new resistivity profile section, we compared the difference between the two. The deleted distances of the data points were 0 m, 2 m, 4 m, 6 m, 8 m, and 10 m from the boundary medium.

This study explored the aspects of the resistivity ratio, electrode spacing, the influence distance of the boundary effect, medium depth, and medium size through parametric variation. In the model, it set that the resistivity of the background soil layer was R1, the resistivity of the boundary medium was *R2*, the section size of the medium was $dyp \times dzp = 2 \times 2$ m^2, the embedding depth was $dep = 2$ m, the spacing of the electrode rods was $dx = 2$, and the length of the survey line was $L = 50$ m. Table 2 is the explanatory table of the boundary effect model parameters.

Table 2. The explanatory table of the boundary effect stratum model parameters.

Model	Influence Factor	*n*	*R1* (ohm-m)	*R2* (ohm-m)	*dx* (m)	*dep* (m)	*dyp* (m)	*dzp* (m)	*L* (m)
Boundary effect model	Resistivity ratio ($n = R2/R1$)	0.01 0.05 0.1 0.25	1000	10 50 100 250	2	2	2	2	50
	Medium size	0.01	1000	10	2	2	2 4	2 4	50
	Embedding depth	0.01	1000	10	2	2 4	2	2	50
	Influence distance	0.01	1000	10	2	2	2	2	50
	Electrode spacing	0.01	1000	10	2 4	2	2	2	50 100

4. Result and Discussion

4.1. Pipeline Stratum Model under 3D Effect

4.1.1. Resistivity Ratio ($n = R2/R1$)

To understand the influence of the geological resistivity ratio of the pipeline stratum to the 3D effect, this study set four different kinds of geological resistivity ratio values ($n = 0.01$, 0.05, 0.1, and 0.25). We fixed the *R1* as 1000 ohm-m, and set the *R2* as 10, 50, 100, and 250 ohm-m, respectively, for the discussion on the influence of the resistivity ratio.

This study analyzed the influence through small pipeline sizes (1.5 m × 2 m), and the survey line L1~L5 of the small-scale pipeline was sequentially 1.5, 0, 3, 6, and 9 m away from the boundary of the pipeline, therefore, we adopted L3 (where the horizontal distance from the pipeline was 3 m) to analyze. The result of its resistivity ratio is shown in Figure 5. When the resistivity ratio was less than 0.05, the mapping phenomenon of the 3D effect could be clearly observed, and when the resistivity ratio was more than 0.1, the 3D effect was gradually nonsignificant.

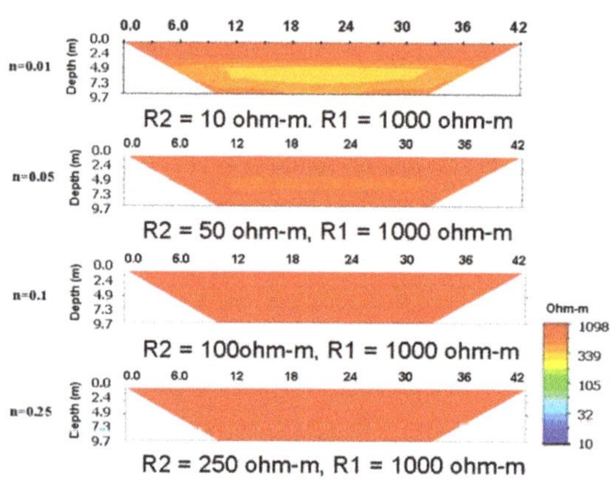

Figure 5. The resistivity ratio and the 3D effect influence on the pipeline stratum model.

4.1.2. Pipeline Size

To understand the influence of the pipeline size on the 3D effect, this study set two kinds of stratum models with different pipeline sizes, where the section size of the pipelines was 1.5 (W) × 2 (D) m and 3 (W) × 4 (D) m, respectively. As seen in Section 4.1.1, when n was more than 0.1, the 3D effect became weaker. Therefore, this study fixed the resistivity ratio to explore the 3D effect under different conditions of pipeline sizes.

Figure 6 shows the results of testing the 2D ERT near two different pipeline sizes. Due to the pipeline sizes, the distances between the L1~L5 lines of the small-scale pipeline and the boundary of the pipeline were 1.5, 0, 3, 6, and 9 m, respectively, and the distances between the L1~L6 lines of the large size pipeline and the boundary of the pipeline were 0 (the left side of the pipeline), 0 (the right side of the pipeline), 3, 6, 9, and 12 m, respectively. In the case of the small pipeline, there was a nonsignificant 3D effect outside the line of the boundary, but in the large size pipeline, line L4 (6 m away from the edge of the pipeline) shows the mapping results of adjacent pipelines. The results show that the influence of the 3D effect increases with the size of the pipeline.

Furthermore, we observed that the mapping depth was deeper than the actual depth of the pipeline. The results showed that the mapping mechanism was not horizontally mapped, and that the mapping depth was related to the distance between the pipeline and the survey line. When the distance was over 6–9 m, the result showed that it was almost unaffected by the 3D effect.

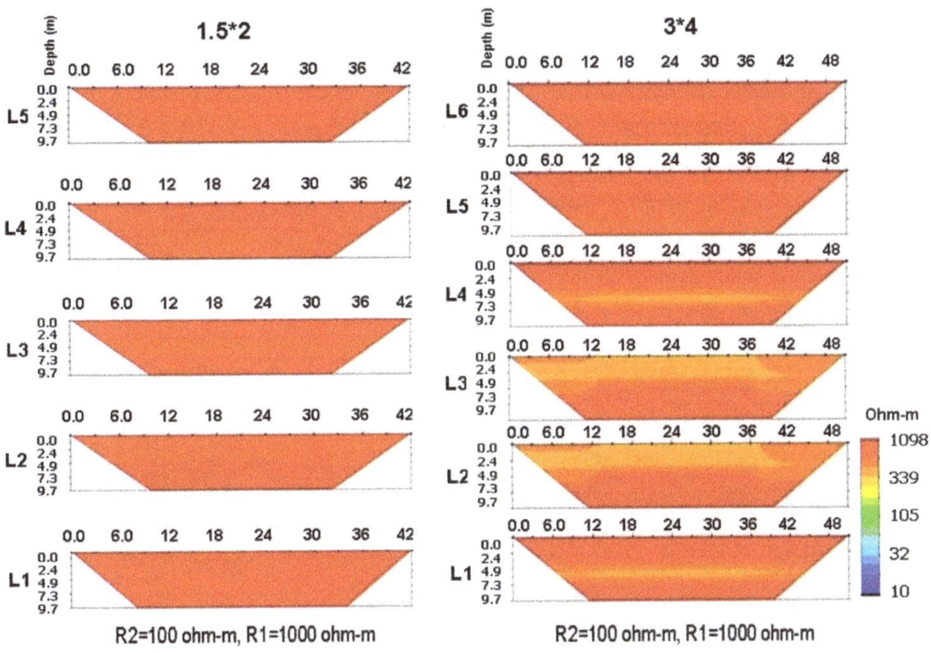

Figure 6. The influence of the pipeline size of the pipeline stratum model on the 3D effect.

4.1.3. Embedding Depth

To understand the influence of the pipeline embedding depth on the 3D effect, this study set two kinds of models with different depths of embedded pipeline. The embedding depth (the top of the pipeline) was 0 m (embedding at the surface) and 4 m, respectively. The n value in the model was set as 0.1, the section size of the pipeline was 1.5 (W) × 2 (D) m, and there were five survey lines.

Figure 7 shows the 2D ERT results of the small pipeline with an embedding depth of 0 m and 4 m. Compared with the L1 lines (the horizontal distance from the buried pipeline was 1.5 m), when the embedding depth was 2 m, we could obviously observe the mapping phenomena, but when the embedding depth increased to 4 m, the 3D effect was not as obvious. This shows that as the depth increased, the influence of the 3D effect weakened, and the influence range seems to be related to the distance from the line to the pipeline, and not the horizontal distance.

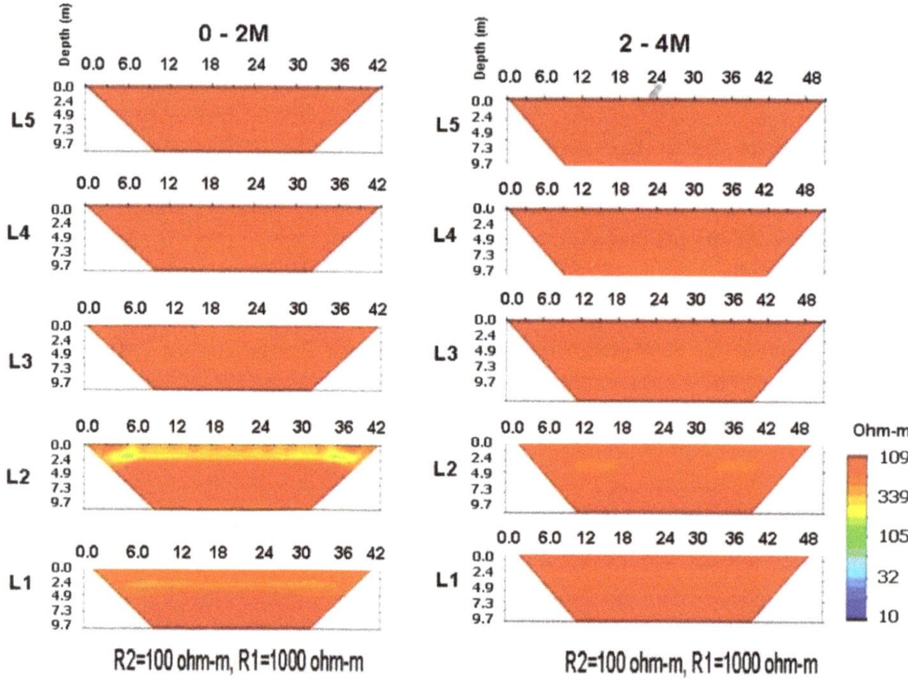

Figure 7. The influence of the pipeline embedding depth in the pipeline stratum model on the 3D effect.

4.1.4. Influence Distance

To find out the possible influence range of the 3D effect in the pipeline stratum, this study used the model that set n as 0.1. The section size of the pipeline was 1.5 (W) × 2 (D) m, $dep = 2$ m, $dy = 3$ m, $dx = 3$ m, $L = 42$ m, and the L1~L5 lines were 1.5, 0, 3, 6, and 9 m away from the boundary of the pipeline.

As shown in the image on the left in Figure 7, the profiles at L1 and L3 had a significant 3D effect, whereas the 3D effect of the L4 line profile gradually became less obvious, and in the L5 line profile, there was no 3D effect. Therefore, the 3D effect influence distance was 6 m, and gradually faded after 6 m.

4.1.5. Electrode Spacing

To understand the relationship between the electrode spacing of the pipeline stratum and the 3D effect, and whether the electrode spacing could be used as the normalized parameter, this study analyzed the electrode spacing parameter, which was 3 m and 1.5 m, respectively. The model set n as 0.1, the section size of the pipeline was 1.5 (W) × 2 (D) m, $dep = 2$ m (the top of the pipeline), $dy = 3$ m, $dx = 1.5$ m or 3 m, and $L = 42$ m.

We used the profiles of L1~L6 to compare the influence situation and influence distance of the 3D effect with different electrode spacings ($dx = 1.5$, 3). As shown in Figure 8, when the electrode spacing was shortened, the spatial resolution improved. However, the influence range of the 3D effect between different electrode spacings was similar; after the distance from the edge of the pipeline increased to

6 m, the 3D effect gradually became inconspicuous. This result shows that it is not appropriate to use electrode spacing as a normalized parameter for the 3D effect because the influence spacing does not decrease as the electrode spacing decreases.

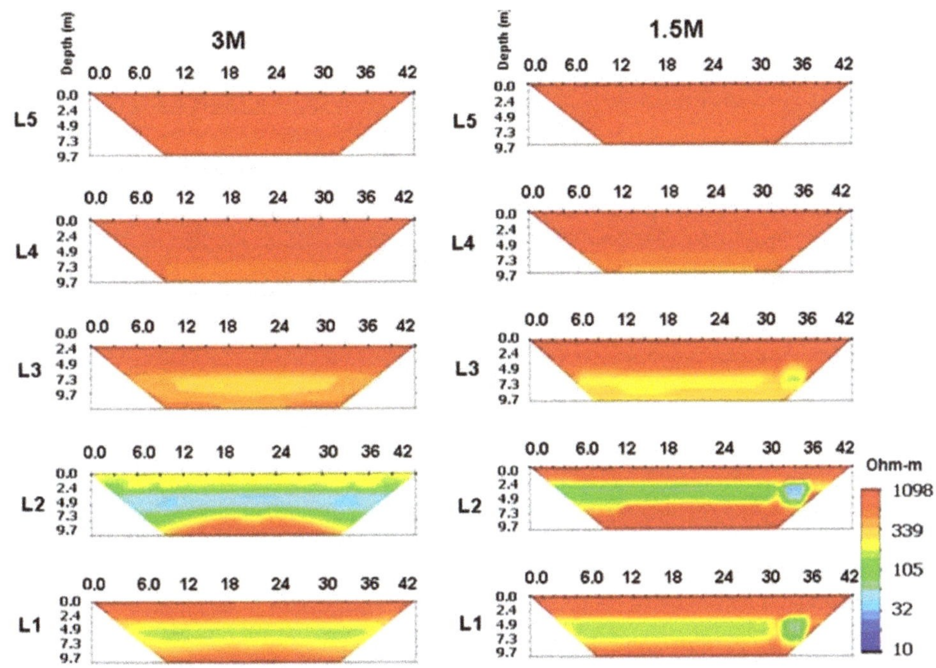

Figure 8. Pipeline stratum model showing the influence between the 3D effect range and electrode spacing.

4.2. Boundary Effect Model

4.2.1. Resistivity Ratio ($n = R2/R1$)

To understand the influence of the boundary effect in different resistivity ratios, this study fixed the background value *R1* as 1000 ohm-m, and assumed that the ratio *n* of *R2* and *R1* were 0.25, 0.1, 0.05, and 0.01, respectively, that is, the model of *R2* resistivity was 250, 100, 50, and 10 ohm-m, respectively.

The results in Figure 9 show that the presence of the medium in line boundary caused a resistivity profile with an unusual resistivity value near the boundary, and the unusual boundary resistivity value was higher than the background value (1000 ohm-m). The study judged that the current encountered the low resistivity medium in the boundary when it flowed through the boundary. This creates a large amount of current that is concentrated in a low-resistivity zone, and causes the resistivity value of the boundary profile to increase abnormally.

In the case of $n < 1$ (*R2/R1*), with the increase in the resistivity ratio value n, the influence of the boundary effect gradually decreased. As shown in Figure 9, there was an abnormal resistivity value zone in the resistivity profile when the resistivity ratio value n was under 0.1, and was not obvious until the value n reached 0.25. That is, when resistivity ratio is under about 0.1, there are concerns about the boundary effect, which is similar to the situation of the 3D effect.

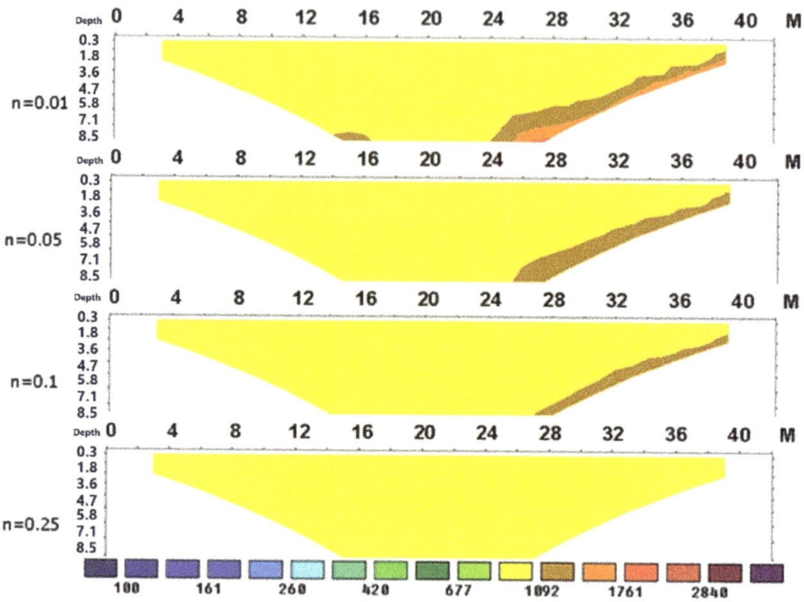

Figure 9. The influence situation of the boundary effect with different resistivity ratios.

4.2.2. Medium Size

To understand the influence of the medium size on the boundary effect, this study fixed the electrode spacing and the medium depth, and used the medium size of 2 × 2 m and 4 × 4 m to explore this further.

The results in Figure 10 show that under the same value n, the addition of the size of the boundary medium does not affect the influence range of the boundary effect; that is, regardless of the size of the boundary medium, the boundary effect is the same, and this situation is different from the 3D effect of the pipeline stratum. It can also be seen in Figure 10 that the influence ranges of the resistivity ratio $n = 0.1$ were the same regardless of the pipe size, and the influence distances were both 8 m.

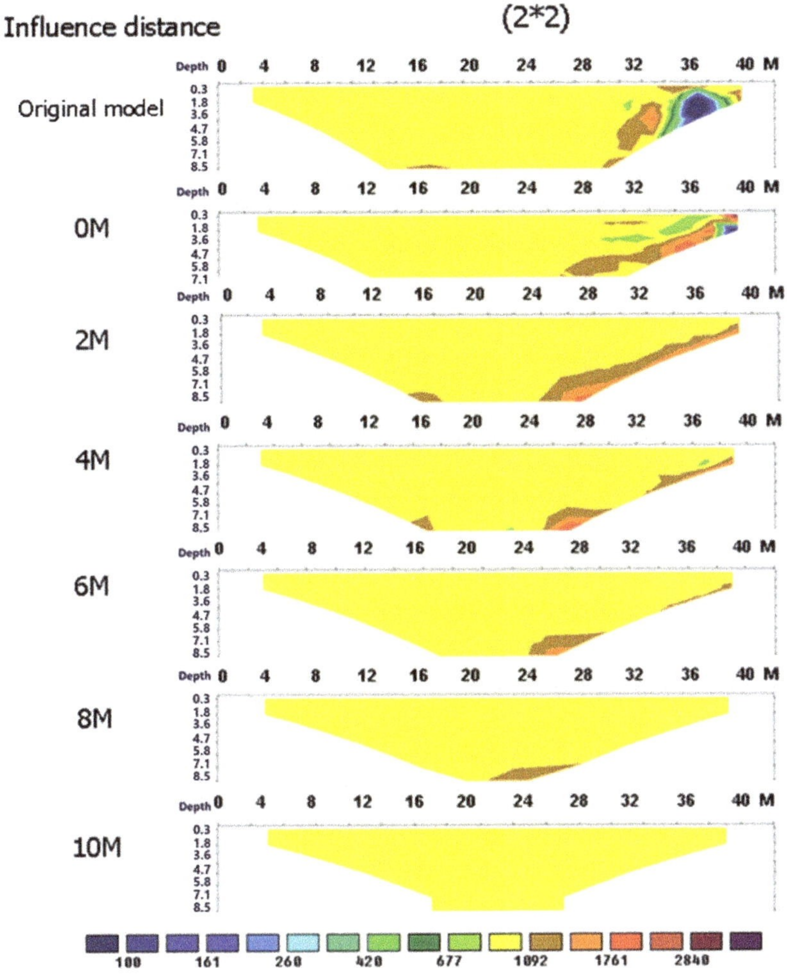

(a)

Figure 10. *Cont.*

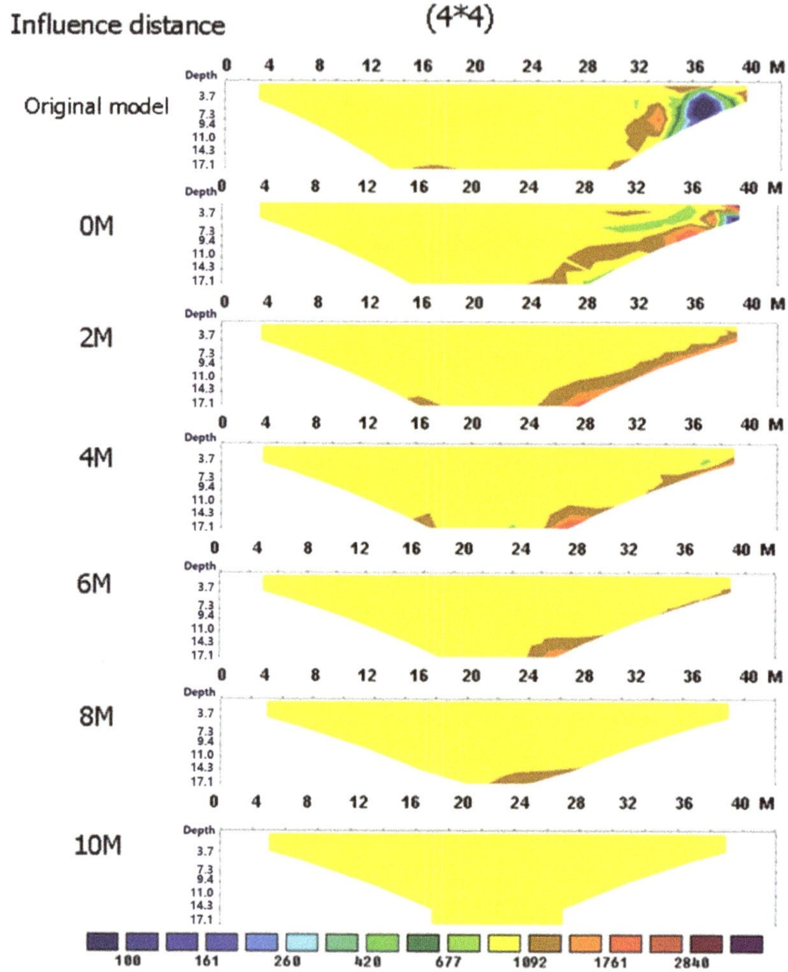

(b)

Figure 10. The influence situation of the boundary effect with different medium sizes. (**a**) medium size 2 × 2; (**b**) medium size 4 × 4.

4.2.3. Embedding Depth

In order to understand the influence of different embedding depths on the boundary effect, this study explored the medium that was 2 m and 4 m under the surface, respectively.

The results in Figure 11 show that if the boundary medium was the same size, the influence distance of the boundary medium under the surface 4 m was farther than the one under the surface at 2 m. That is, the deeper the boundary medium, the larger the influence of the boundary effect, which is a different situation to that of the 3D effect of the pipeline boundary. It can be seen in Figure 11 that if the resistivity ratio value n was 1 when *dep* = 2 m, the influence distance of the boundary effect was 8 m, and when *dep* = 4 m, the influence distance of the boundary effect was 10 m. This result can be seen by the data point of the apparent resistivity in the model; when the medium below the surface is deeper, the deeper data point will be affected, which will cause the influence range of the boundary effect to become wider.

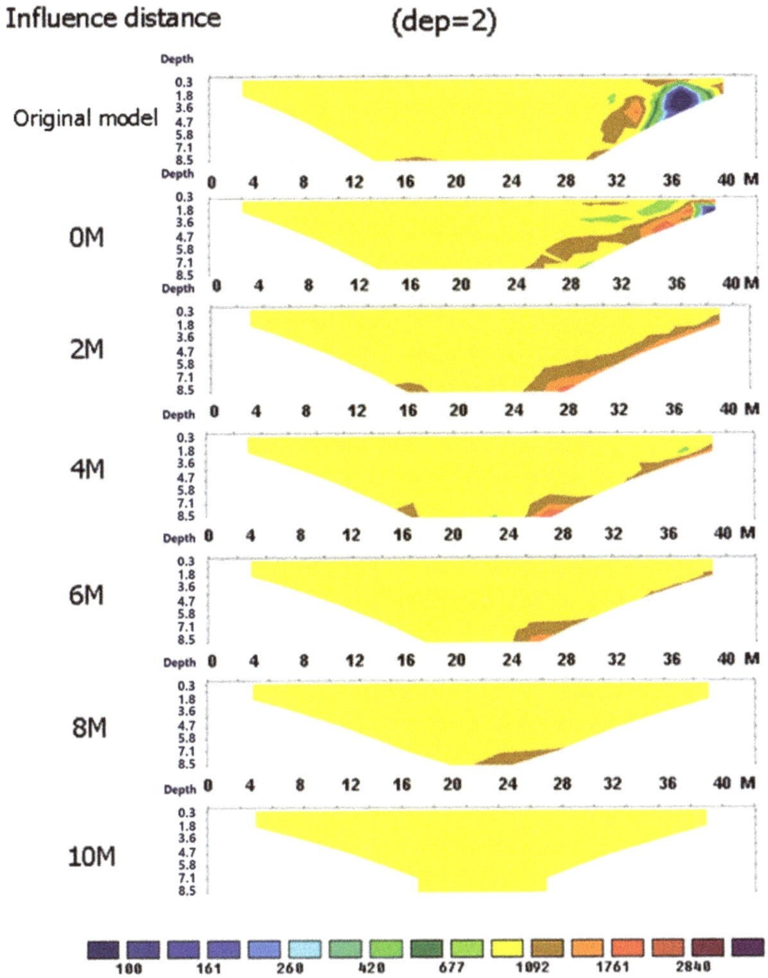

(a)

Figure 11. *Cont.*

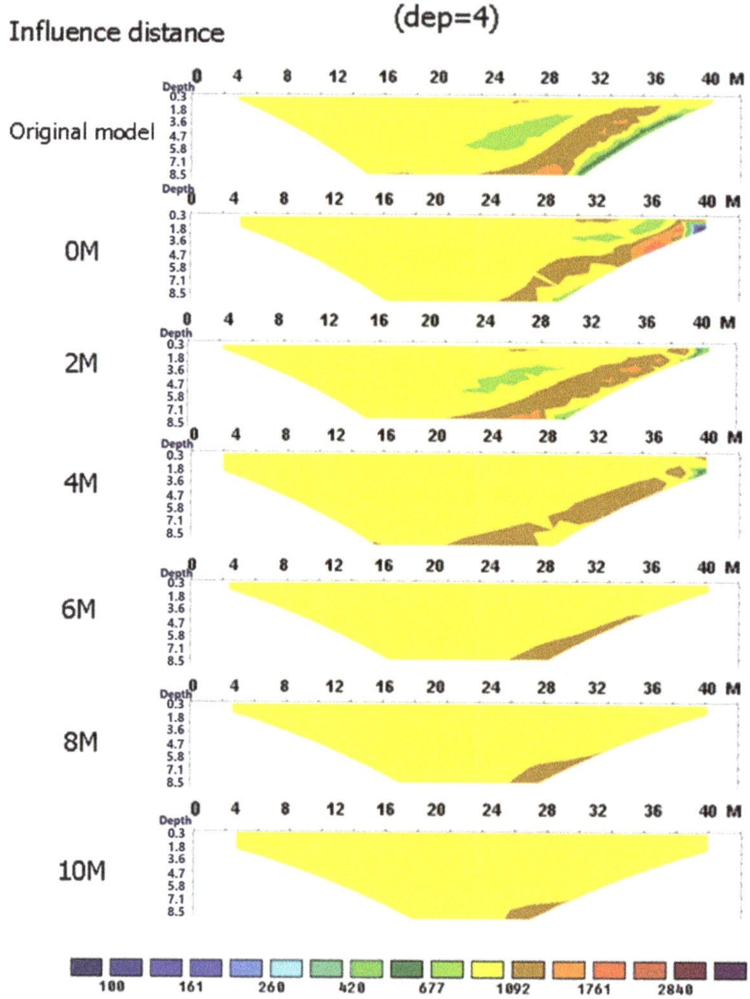

(b)

Figure 11. The influence of the boundary effect with different medium depths. (**a**) medium depth 2 m; (**b**) medium depth 4 m.

4.2.4. Influence Distance

To understand the influence distance of the boundary effect, this study set the resistivity ratio $n = 0.1$ and electrode spacing $dx = 2$. These results are shown in the left image in Figure 12, where the resistivity profile obviously had an abnormal resistivity zone within 6 m from the boundary medium, which was obviously affected by the boundary effect. However, after 8 m away from the boundary medium, it was almost unaffected by the boundary effect. This result is similar to the influence distance of the 3D effect.

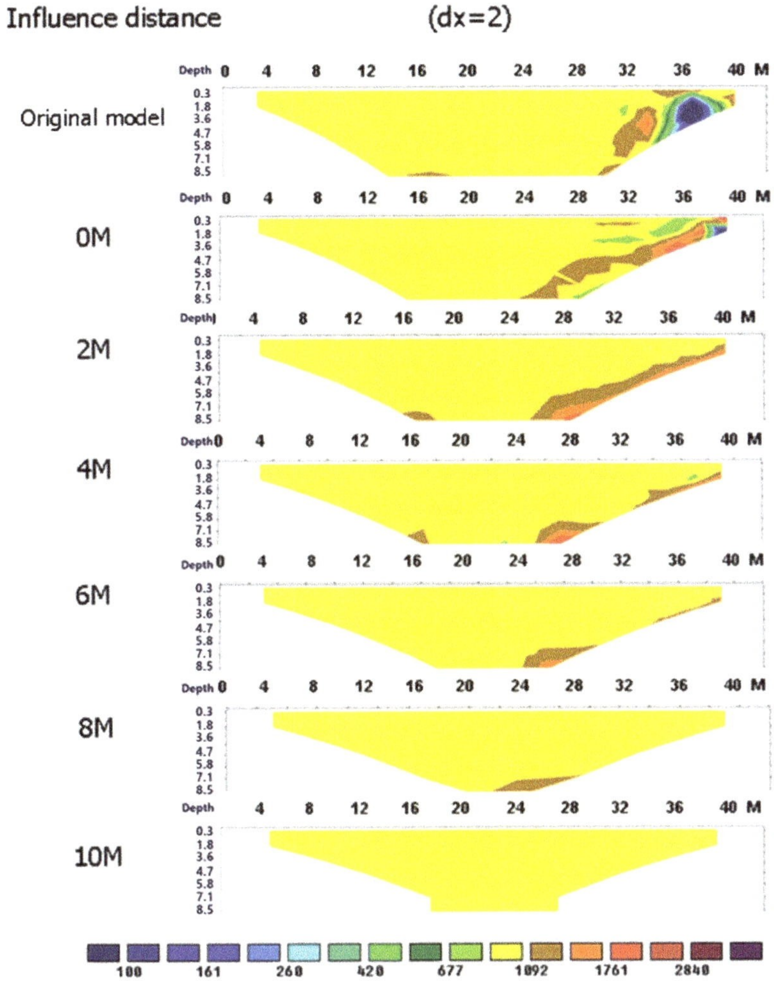

(a)

Figure 12. *Cont.*

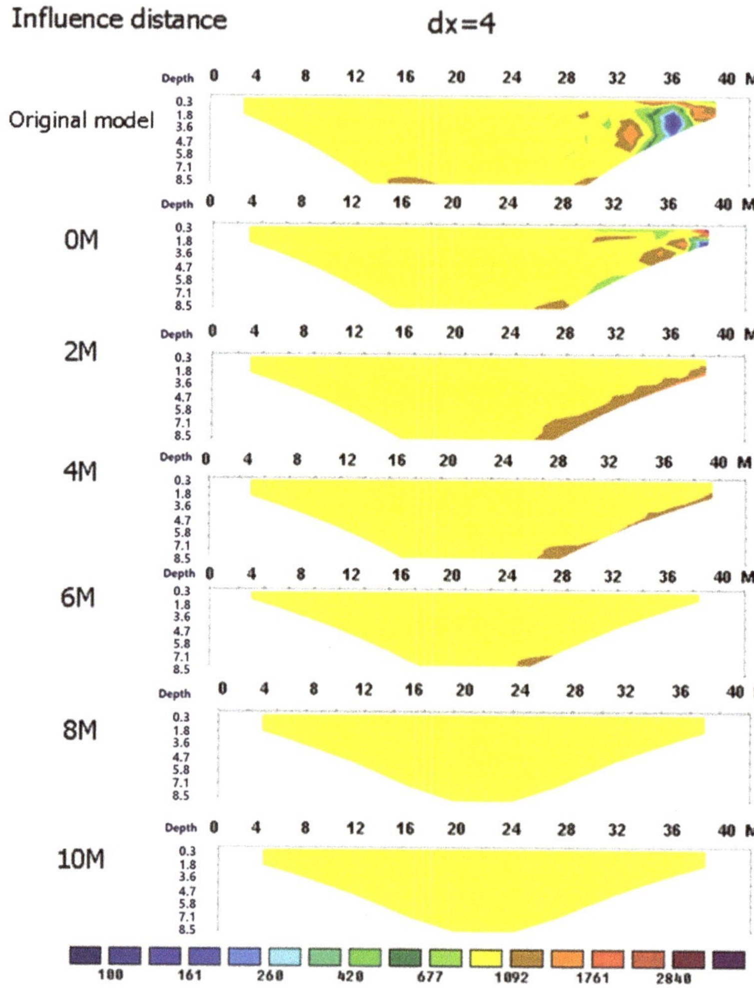

(b)

Figure 12. The influence situation of the boundary effect with different electrode spacing. (**a**) electrode spacing 2 m; (**b**) electrode spacing 4 m.

4.2.5. Electrode Spacing

To understand the influence between the electrode spacing and the boundary effect, and if the electrode spacing could be a normalization parameter, or whether its effect was a fixed distance regardless of the electrode rod spacing, this study fixed the medium depth and size and set the electrode spacing as 2 m and 4 m, respectively.

When the resistivity ratio value n was 0.1 and the electrode spacing dx was 2, the influence distance of the boundary effect was 8 m; moreover, when the electrode spacing dx was 4, the influence distance of the boundary effect was also 8 m. Although the use of a smaller electrode spacing had better resolution, the influence distances were the same. Therefore, it is not appropriate to use the electrode spacing as the normalization parameter of the influence space of the boundary effect as the influence distance does not become smaller as the electrode spacing becomes smaller. This result is similar to the 3D effect of the pipeline stratum, where neither were affected by the electrode spacing.

4.3. Discussion

2D ERT assumes that the geological formation is a 2D semi-infinite space. However, in the real state of geological formation, the electric current flows in the 3D (X, Y, Z) direction, therefore, the object in the non-2D section has a certain degree of disturbance to the ground resistivity electric field, thus causing some irregular resistivity and noise on the 2D profile [8], which is called the 3D effect. Many scholars have adopted indoor sandbox experiments to conduct 2D ERT. As a result, a space with boundary constraints is formed. The current transfer is affected by the boundary, and causes some irregular resistivity and noise on the 2D profile, which is called the boundary effect. Most scholars will often choose to evade the 3D effect and boundary effect when performing 2D ERT. Few scholars had made in-depth discussions on the 3D effect and boundary effect. Mei, Xing-Tai used a homogeneous thickness model made from dough to explore the indoor test method and criteria for 2D ERT where. The results of this study and that of Mei, Xing-Tai (2001) both found that the 2D resistivity profile had a boundary effect. Mei, Xing-Tai found that the sandbox experiment needed to have twice the distance of spread to avoid the boundary effect. This study found that the boundary effect had a distance of about 6–8 m. Yang and Lagmanson used a numerical simulation to build a model with three different resistivity blocks, and placed a 2D and 3D survey line on the surface of the model. The results showed that the 2D resistivity profile was mapped by surrounding resistivity blocks and exhibited an irregular resistivity distribution; however, the 3D resistivity profile was not affected [8]. The results of this study and that by Yang and Lagmanson (2006) both showed that the 2D resistivity profile had a 3D effect. After deeper analysis, the study found that the influence distance was 8–6 m, and that the three-dimensional effect was not obvious after the distance limit of 8 m. Lin et al. conducted 2D ERT at the Hsin-Shan Earth Dam in Taiwan where during the measurements, it was found to be affected by the changes of the nearby stratum and geocenter as well as the curtain grouting wall of the Hsin-Shan Earth Dam, which was at 45 degrees with the survey line. As the 2D resistivity profile had a 3D effect, the authors established the 3D forward modeling of the Hsin-Shan Earth Dam, and compared the differences between the 2D resistivity profile after inverse computing and the Hsin-Shan Earth Dam model. The results showed that the 2D resistivity profile was indeed affected by the 3D effect and had an irregular resistivity distribution [10]. Both the results of this study and Lin et al. (2013) found that the 2D ERT had a 3D effect. However, the results of this study explored the effects of various influencing factors on the 3D effect and boundary effect in depth, and should be avoided as much as possible in future detection.

In summary, most of the studies that have explored the 3D effect and boundary effect have mostly been to discover and confirm the existence of the 3D effect and boundary effect, or to explore the effects of the 3D effect and boundary effect on a single parameter. As a result, this study explored the effects of the 3D effect and boundary effect on 2D ERT through the parameters of the resistivity ratio, pipeline size, embedding depth, influence distance, and electrode spacing under different conditions and summarized these changes as conclusions and discussed them in detail. In this study, the 3D effect and boundary effect of the 2D resistivity profile were explored through numerical simulation. From the simulation results, good results were obtained, and the relevant results can be used to provide a reference for future detection.

5. Conclusions and Suggestions

ERT can provide the stratum 2D or even 3D resistivity profile to further understand the situation of stratigraphic change, but it is a big challenge to engineer the data interpretation of the ERT spatial resolution and testing result. One important phenomenon in the 3D effect that was observed whilst exploring the pipeline stratum model was the mapping phenomenon of the 3D effect of the pipeline stratum, which will produce an illusion of high and low resistivity in the 2D resistivity profile.

The boundary effect will cause the resistivity value of the 2D profile boundary to abnormally increase, reduce resolution, and increase interpretation error. Through the model established in this study, it was found that there was roughly similar influence trend in the parameter analysis of the 3D and boundary effects. The following are suggestions for future testing.

1. Under different resistivity ratios (n), when the n value increases (that is, the difference in resistivity between the formation and the medium becomes smaller), the influence ranges of the 3D and boundary effects both decrease.
2. The 3D and boundary effects have a similar influence range, with an influence distance of 6~8 m.
3. Different electrode spacings showed that it is not appropriate to use electrode spacing as a normalization parameter because the influence distance does not increase as the electrode spacing increases.
4. With different medium sizes, an increase in the pipeline size will increase the range of the 3D effect, but the boundary effect is not affected by the size of the medium.
5. At different embedding depths, the deeper the buried depth, the smaller the 3D effect of the pipeline, but the greater the boundary effect of the medium. This is because the deeper the boundary medium, the more the resistivity changes at the deeper boundary profile.

Currently, the scope of this study is only to establish a numerical model. In the future, we will try to confirm the results of the numerical simulation through in situ testing.

Author Contributions: Conceptualization, C.-P.L.; Methodology, C.-P.L.; Modeling Y.-C.H.; Software, Y.-C.H. and C.-T.L.; Data analysis Y.-C.H.; Conclusions, C.-P.L. and Y.-C.H.; Electrology consultation, C.-T.L. and K.-W.W.

Funding: This study has no funding.

Conflicts of Interest: The authors declare no conflict of interest.

References

1. Meads, N.L.; Bentley, L.R.; Mendoza, C.A. Application of electrical resistivity imaging to the development of a geologic model for a proposed Edmonton landfill site. *Can. Geotech. J.* **2003**, *40*, 551–558.
2. Schoor, M. Detection of sinkholes using 2D electrical resistivity imaging. *J. Appl. Geophys.* **2002**, *50*, 393–399.
3. Dahlin, T.; Bernstone, C.; Loke, M.H. A 3-D resistivity investigation of a contaminated site at Lernacken. *Sweden. Geophys.* **2002**, *67*, 1692–1700.
4. Demanet, D.; Pirard, E.; Renardy, F.; Jongmans, D. Application and processing of geophysical images for mapping faults. *Comput. Geosci.* **2001**, *27*, 1031–1037.
5. Dahlin, T. The development of DC resistivity imaging techniques. *Comput. Geosci.* **2001**, *27*, 1019–1029.
6. Nguyen, F.T.; Garambois, S.; Jongmans, D.; Pirarda, E.; Loke, M.H. Image processing of 2D resistivity data for imaging faults. *J. Appl. Geophys.* **2005**, *57*, 260–277.
7. Chen, T.T.; Hung, Y.C.; Hsueh, M.W.; Yeh, Y.H.; Weng, K.W. Evaluating the Application of Electrical Resistivity Tomography for Investigating Seawater Intrusion. *Electronics* **2018**, *7*, 107.
8. Yang, X.; Lagmanson, M. Comparison of 2D and 3D Electrical Resistivity Imaging Methods. *Symp. Appl. Geophys. Eng. Environ. Probl.* **2006**, 585–594. [CrossRef]
9. Dahlin, T. 2D resistivity surveying for environmental and engineering applications. *First Break* **1996**, *14*, 275–283.
10. Lin, C.P.; Hung, Y.C.; Wu, P.L.; Yu, Z.H. Performance of 2D ERT in Investigation of Abnormal Seepage: A Case Study at the Hsin-Shan Earth Dam in Taiwan. *J. Environ. Eng. Geophys.* **2013**, *13*, 101–112.
11. Tabbagh, J.; Samouëlian, A.; Tabbagh, A.; Cousin, I. Numerical modelling of direct current electrical resistivity for the characterisation of cracks in soils. *J. Appl. Geophys.* **2007**, *62*, 313–323.
12. Kowalczyk, S.; Maślakowski, M.; Tucholka, P. Determination of the correlation between the electrical resistivity of non-cohesive soils and the degree of compaction. *J. Appl. Geophys.* **2014**, *110*, 43–50.
13. Sentenac, P.; Zielinski, M. Clay fine fissuring monitoring using miniature geo-electrical resistivity arrays. *Environ. Earth Sci.* **2009**, *59*, 205–214.

14. Jones, G.; Zielinski, M.; Sentenac, P. Mapping desiccation fissures using 3-D electrical resistivity tomography. *J. Appl. Geophys.* **2012**, *84*, 39–51.
15. Berternann, D.; Schwarz, H. Bulk density and water content-dependent electrical resistivity analyses of different soil classes on a laboratory scale. *Environ. Earth Sci.* **2018**, *77*, 570.
16. Hashem, R.R.; Mohammad, K.H.; Mohammad, R.S.K.; Mohammad, R.G.N. Electrical resistivity method for water content and compaction evaluation, a laboratory test on construction material. *J. Appl. Geophys.* **2019**, *168*, 49–58.
17. Mei, X.T. 2-D Preliminary Analysis of RIP. Ph.D. Theses, National Taiwan University Civil Engineering Institute Theses and Dissertations, Taipei, Taiwan, December 2012.
18. Koefoed, O. *Geosounding Principles*; Elsevier Scientific Publishing Company: Oakland, CA, USA, 1979.
19. Oldenburg, D.; Li, Y. Estimating the depth of investigation in dc resistivity and IP surveys. *Geophysics* **1999**, *64*, 403–416.
20. *EarthImager (Computer software)*; International Parkway Virginal Beach, AGI: Austin, TX, USA, 2009.

applied sciences

MDPI

Article

Area-Efficient FFT Kernel with Improved Use of GI for Multistandard MIMO-OFDM Applications

Song-Nien Tang [1,*] and Yuan-Ho Chen [2,*]

1 Information and Computer Engineering Department, Chung Yuan Christian University, Taoyuan 32023, Taiwan

2 Department of Electronics Engineering, Chang Gung University, and Institute for Radiological Research, Chang Gung University/Chang Gung Memorial Hospital, Taoyuan 33302, Taiwan

* Correspondence: sntang@ice.cycu.edu.tw (S.-N.T.); chenyh@mail.cgu.edu.tw (Y.-H.C.); Tel.: +886-3-265-4731 (S.-N.T.)

Received: 17 May 2019; Accepted: 16 July 2019; Published: 18 July 2019

Abstract: This study presents a fast Fourier transform (FFT) kernel for multistandard applications, which employ multiple-input, multiple-output orthogonal frequency-division multiplexing (MIMO-OFDM). The proposed design uses a mixed-radix, mixed-multipath delay-feedback (MRM^{2}DF) structure, which enables 4/5/6-stream 64/128-point FFT. This approach allows the effective usage of guard intervals (GI) in conjunction with a novel resource-sharing scheme to improve area efficiency. An area-reduced constant multiplication unit and sorting buffer with minimal memory size further reduced an area overhead. A test chip was designed using UMC 90-nm technology, and was evaluated through post-layout simulation. The proposed design outperformed previous works in terms of the throughput per area.

Keywords: fast Fourier transform; FFT; kernel; MIMO; OFDM; multistandard

1. Introduction

With the development of household and industrial applications, wireless data accesses between devices and users have become a concern on demand. Among various wireless communication schemes, multiple-input, multiple-output orthogonal frequency-division multiplexing (MIMO-OFDM) technology [1] are superior in transmission performances and are widely employed in numerous standards, such as WiFi (IEEE 802.11n/ac/ax), Long Term Evolution (LTE), and advanced 5G New Radio (NR) [2–5]. In practical applications, to access data from various devices, the user side must structure MIMO-OFDM communication according to the standard employed by the individual device. Therefore, developing a user-side transceiver, which can perform a multistandard MIMO-OFDM transmission for different devices, is useful (Figure 1).

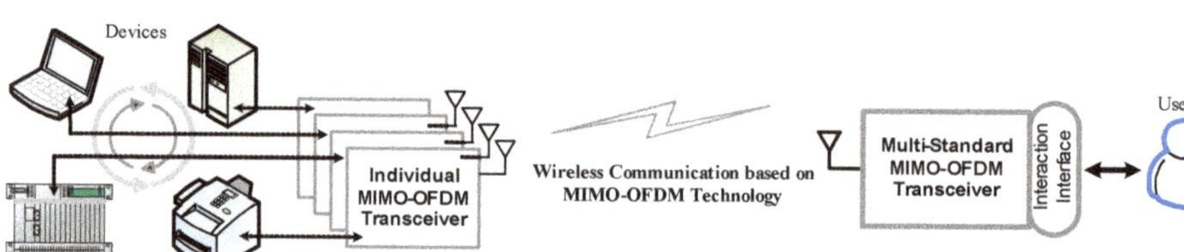

Figure 1. Scheme illustration of multistandard multiple-input, multiple-output orthogonal frequency-division multiplexing (MIMO-OFDM) data access applications.

For all MIMO-OFDM communication between users and devices, Figure 2 shows the block diagram of a conventional $M \times M$ MIMO-OFDM transmission model. A transmitter (TX) sends M

streams of data symbols through M antennas. To reduce intersymbol interference (ISI) in OFDM communication, a guard interval (GI) was applied to the symbols by copying the number of samples from a symbol end to a symbol head. A receiver (RX) then uses M antennas to receive signals, which was obtained through the $M \times M$ MIMO channel. Figure 2 shows that the RX part contains M sets of radio frequencies (RF), analog-to-digital convertors (ADC), fast Fourier transforms (FFTs) and the following equalization and demodulation blocks. For the given N samples in an OFDM symbol, an $M \times M$ MIMO-OFDM receiver requires M-stream N-point FFT operations. The TX requires to perform M-stream N-point inverse FFT (IFFT) operations. Therefore, a set of M-stream N-point FFT (IFFT) processors are essential components in the MIMO-OFDM system. In general, the maximum M and N values for standard MIMO-OFDM FFT are 8 and 2048, respectively [6,7], and they can be higher in 5G applications [5]. Therefore, the cost of hardware and operation throughput for the required FFT/IFFT processors are significant design concerns. Further considering the FFT/IFFT design for multistandard MIMO-OFDM applications, implementing an effective scheme to develop the multimode hardware configuration is a difficult task. In this study, an FFT kernel design, which could be structured as a base module for reconfigurable FFT/IFFT processors, was proposed in a multistandard MIMO-OFDM system. The proposed FFT kernel could perform 4/5/6-stream 64/128-point FFT operations by efficiently using GI duration with resource-sharing and operation-rescheduling schemes. Therefore, the presented FFT kernel supports area-efficient and high-throughput development for multistandard MIMO-OFDM FFT/IFFT processors. The remainder of the paper is organized as follows. Section 2 outlines the design considerations for the FFT design and literature. Section 3 presents the hardware architecture of the proposed FFT kernel. Section 4 reports the implementation and comparison of the proposed FFT kernel design. Finally, conclusions are presented in Section 5.

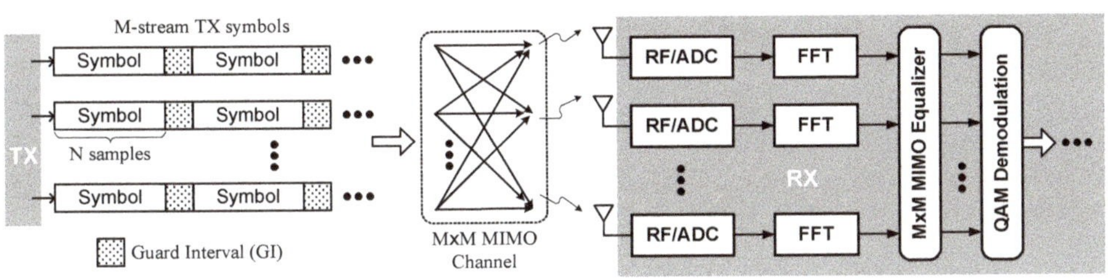

Figure 2. Block diagram for the conventional $M \times M$ MIMO-OFDM transmission model.

2. FFT Design Considerations

2.1. Related Works for MIMO-OFDM FFT

Various hardware architectures have been developed to perform the FFT algorithm. Among these architectures, the memory-based [8] and pipelined [9] architecture are the two primary categories. The memory-based FFT configuration is structured based on the employment of one or several processing elements (PEs) in cooperation with the memory modules. This memory-based structure is suitable for flexible FFT operations, which provide various FFT streams or operation points (i.e., FFT length) [10–12]. However, this approach is not appropriate for high-throughput or low-latency FFT operations when the operability of PEs or memory bandwidths is limited [13–15]. By using flexible FFT computation (length or streams), numerous memory-based FFT designs have been presented for MIMO-OFDM applications [15–17]. However, most of these approaches employ low-radix PEs (i.e., radix-r, where r is ≤ 16), which limits throughput performance. The pipelined FFT architecture is applied to increase operation throughput and latency by using hardware resources [9]. Moreover, two conventional pipelined FFT categories are multipath delay communicator (MDC) [9,18] and single-path delay-feedback (SDF) architectures [9,19].

The MDC structure has the features of multipath (i.e., parallel) feedforward operations performed using switch control with first in first out (FIFO) memory. By effectively using the parallel data paths and operation units, the multipath processing feature of MDC can be applied to the FFT computation of multiple streams [20]. Therefore, several FFT designs based on MDC structures have been presented for MIMO-OFDM applications [20–23]. Figure 3 shows the block diagram for the conventional M-path MDC architecture applied to MIMO-OFDM FFT of M streams, where the feedforward multipath data is processed using switch blocks, FIFOs, butterfly units and multipliers. The SDF architecture provides a feedback path for FIFOs to efficiently manage the butterfly operation data at each pipeline stage [9]. To enable parallel processing for SDF configurations, the extended multipath delay-feedback (MDF) architecture [24,25] was proposed. Because the MDF scheme can process multiple input streams, MDF architectures have been researched for required multistream FFT operations in MIMO-OFDM systems [7,25–29]. On the basis of the MDF structure using the radix-2^k algorithm (where k is a positive integer) [30], an N-point FFT unit has $\log_2(N)$ radix-2 operation stages involving the feedback-pathed FIFO and butterfly 2 (BF2) units [9,30]. Figure 4 shows a conventional M-path MDF FFT structure for $M \times M$ MIMO-OFDM applications. Furthermore, some previous studies [21,25,28] discuss the schemes of hardware cost reduction for multipliers located at multiple paths (Figures 3 and 4). In general, these MDC or MDF approaches [20–29] enable multipath/parallel operations associated with multiple FFT streams (i.e., the number of data paths and FFT streams is equal).

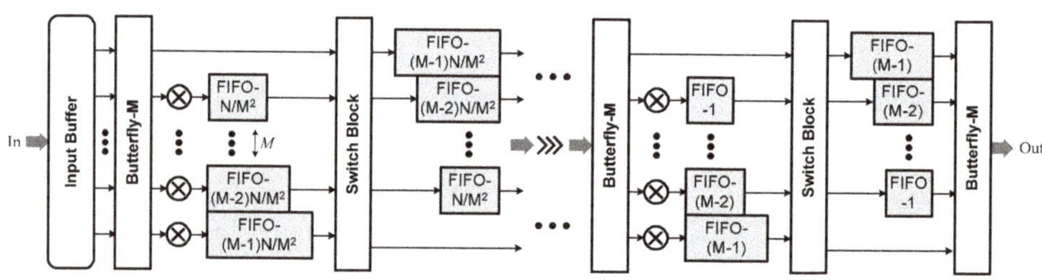

Figure 3. Block diagram of conventional M-path MDC fast Fourier transform (FFT) architecture for M-stream MIMO-OFDM.

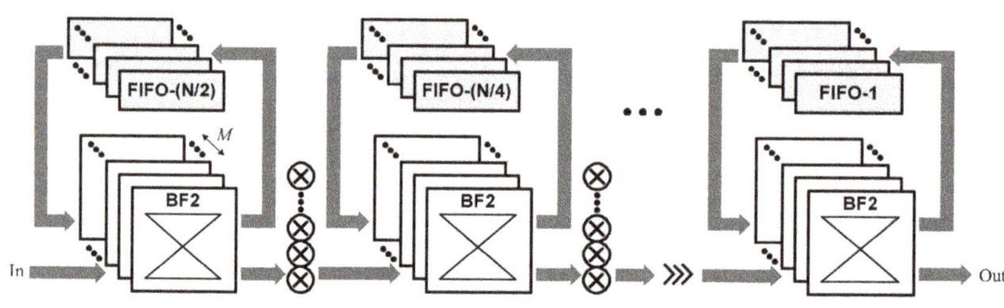

Figure 4. Block diagram of conventional M-path MDF FFT architecture for M-stream MIMO-OFDM.

2.2. *Kernel-Based FFT for Multiple Standards*

Most MIMO-OFDM standards have variable FFT lengths and stream levels, and thus, the corresponding MDC or MDF FFT designs generally employ a reconfigurable structure to support various FFT operation modes specified at the aimed specification [7,20,22]. In addition, considerable research has been conducted on advanced restructure schemes for multimode FFT processors that support multiple MIMO-OFDM standards [28,29]. For such multimode/multistandard MIMO-OFDM FFT applications, the aforementioned studies have employed tailored reconfiguration methods to enable multimode FFT operations based on their individual MDC/MDF structures. Therefore, the aforementioned approaches lack the design flexibility and convenience to develop the multistandard MIMO-OFDM FFT architecture.

However, among the mainstream MIMO-OFDM standards (WiFi, LTE, and 5G applications), 1- to 4-stream 128/64-point FFT is implemented with a base operation mode [20–23,25–28]. This is one approach, using which a common optimized four-stream 128/64-point FFT kernel can be developed as a base module to construct specified M-stream N-point FFT/IFFT processors. For example, in the MDF FFT architecture, Figure 5 presents the aforementioned scheme. Considering M-stream N-point FFT computation ($M > 4$; $N > 128$, and N is to the power of 2), each N-point FFT can be performed using N'-point FFT in conjunction with 128-point FFT based on the radix-N'/128 FFT algorithm, where N' is equal to $N/128$, and M-stream FFT can be implemented in $\lceil M/4 \rceil$ sets of four-stream FFT; $\lceil . \rceil$ denotes a ceiling operation. Figure 5 shows a four-stream 128/64-point FFT kernel prepared first as a common base module. This kernel module was further integrated with a front four-stream N'-point FFT computation unit to complete a set of four-stream N-point FFT. Such configuration (FFT kernel and N'-point FFT unit) can be extended to $\lceil M/4 \rceil$ sets, and complete M-stream N-point FFT operations can be implemented. For hardware modularization, kernel-based FFT configurations are more efficient and flexible when applied to scalable M-stream N-point FFTs for multiple MIMO-OFDM standards. Figure 5 presents a configuration which can be further modified and extended to integrate a conventional FFT kernel with MDC or memory-based units for target FFT computation. Considering four-path 128/64-point FFT kernel designs, employing the available FFT processors, which can perform the same FFT specification, is an efficient approach [21,25]. Nevertheless, for system optimization, specific development of a hardware-efficient FFT kernel as a common module can be advantageous. In this study, the target FFT kernel module was designed using a modified MDF structure as support. In addition to the original four-stream 128/64-point FFT operations, the proposed FFT kernel used GI duration to enable five- and six-stream FFT operations, and thereby enhance the overall throughput with an improved area efficiency. The concept for GI utilization for the FFT design has been mentioned in [31] for the purpose of improving the operation latency. In our design, we further utilized the GI in common with a resource-sharing scheme to increase the operation throughput. The details of the proposed design are discussed in Sections 3 and 4.

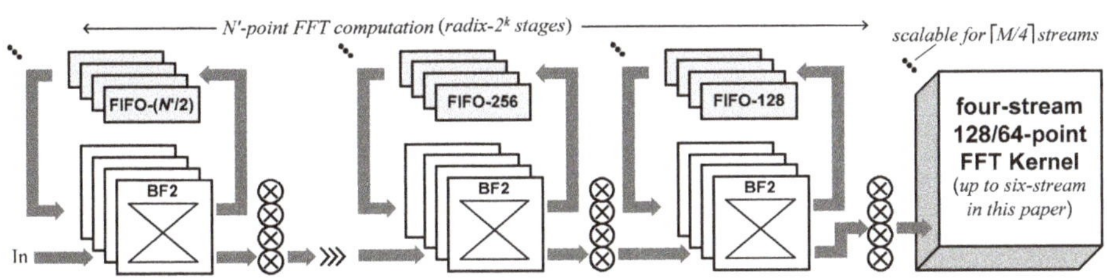

Figure 5. Block diagram showing the hardware architecture of M-stream N-point FFT processor using a. four-stream 128-point FFT kernel.

3. Proposed FFT Kernel Architecture

3.1. Algorithm and Architecture Overview

In conventional MDC/MDF FFT schemes [20–26], the number of paths is equal to the number of FFT streams. By contrast, a mixed-radix, mixed-multipath delay-feedback (MRM^{2}DF) architecture was developed, which operated on a radix-dependent number of paths. For 128-point FFT operations, the MRM^{2}DF design operated based on the mixed radix-2/8/8 algorithm, which can be described using Equations (1)–(5). A 128-point discrete Fourier transform (DFT) of a time domain sequence x(n), was defined as Equation (1), where X(k) and $W_{128}^{nk} = \exp(-j2\pi nk/128)$ are the DFT results and twiddle factor, respectively. Moreover, n and k are represented using Equation (2), and Equation (1) can be derived using Equation (3), a two-step computation based on radix-2 and radix-64 operations represented as BF2 and 64-point DFT, respectively. Between BF2 and 64-point DFT operations,

128-based (W128) twiddle factors must be multiplied with the BF2 output. Furthermore, the 64-point DFT operation in Equation (3) can be further decomposed into the radix-8/8 operation.

By using Equation (4), a 64-point DFT can be derived as Equation (5), where two-stage radix-8 operations are performed. The two radix-8 stages correspond with the first and second butterfly 8 (BF8) operations, respectively. Similarly, 64-based (W64) twiddle factor multiplication is required between the first and second BF8 operations. For 64-point FFT operations, the MRM2DF design executes the radix-8/8 algorithm as Equation (5).

$$X(k) = \sum_{n=0}^{127} x(n)\, W_{128}^{nk},\ k = 0, 1, \ldots, 127 \tag{1}$$

$$\begin{cases} n = 64n_1 + n_2, n_1 = 0, 1; n_2 = 0, 1, \ldots, 63 \\ k = k_1 + 2k_2,\ k_1 = 0, 1; k_2 = 0, 1, \ldots, 63 \end{cases} \tag{2}$$

$$\begin{aligned} X(2k_2 + k_1) &= \sum_{n_2=0}^{63} \sum_{n_1=0}^{1} x(64n_1 + n_2) W_{128}^{(64n_1+n_2)(2k_2+k_1)} \\ &= \sum_{n_2=0}^{63} \left\{ \underbrace{\sum_{n_1=0}^{1} x(64n_1 + n_2)\, W_2^{n_1k_1}\, W_{128}^{n_2k_1}}_{\text{BF2 (radix-2) operation}} \right\} W_{64}^{n_2k_2} = \underbrace{\sum_{n_2=0}^{63} \text{BF2}\,(k_1, n_2)\, W_{64}^{n_2k_2}}_{\text{64-point DFT (radix-64)}} \end{aligned} \tag{3}$$

$$\begin{cases} n_2 = 8\alpha_1 + \alpha_2,\ \alpha_1\, \&\, \alpha_2 = 0, 1, \ldots, 7 \\ k_2 = \beta_1 + 8\beta_2,\ \beta_1\, \&\, \beta_2 = 0, 1, \ldots, 7 \end{cases} \tag{4}$$

$$\begin{aligned} X(2(\beta_1 + 8\beta_2) + k_1) &= \sum_{\alpha_2=0}^{7} \sum_{\alpha_1=0}^{7} \text{BF2}\,(k_1,\, 8\alpha_1 + \alpha_2)\, W_{64}^{(8\alpha_1+\alpha_2)(\beta_1+8\beta_2)} \\ &= \sum_{\alpha_2=0}^{7} \underbrace{\left\{ \underbrace{\left[\sum_{\alpha_1=0}^{7} \text{BF2}(k_1,\, 8\alpha_1 + \alpha_2)\, W_8^{\alpha_1\beta_1} \right]}_{\text{1st BF8 (radix-8) operation}} W_{64}^{\alpha_2\beta_1} \right\} W_8^{\alpha_2\beta_2}}_{\text{2nd BF8 (radix-8) operation}} \end{aligned} \tag{5}$$

Corresponding with the radix-2/8/8 algorithm, the proposed MRM^{2}DF kernel was operated with hybrid path configurations (i.e., mixed-multipath). Hardware units associated with the radix-2 (BF2) computation allowed operations on at most six data paths (i.e., 4/5/6 paths) based on the number of streams. By contrast, radix-8 (BF8) hardware units allowed operations on eight paths corresponding to the number of radices. A shared hardware module was employed to perform W_{128} and W_{64} twiddle factor multiplication in six and eight paths, respectively. When performing 64-point FFT, the MRM^{2}DF kernel performed only eight-path operations based on the radix-8/8 algorithm. The features of the proposed MRM^{2}DF structure are as follows:

(i) In mixed-multipath operations based on streams or radix-8, GI duration can be employed to conduct 128-point FFT operations in up to six streams.
(ii) Resource sharing is applied to multipath radix-2/8/8 operations by using an area-reduced constant multiplication unit.
(iii) The required memory size (i.e., the number of stored data elements) are maintained at a modest level by using a sophisticated sorting scheme applied to the buffer.

(iv) Points (i)–(iii) allow significant improvements in area efficiency compared with available designs that support multipath 128-point FFT (e.g., [21,25]).

Figure 6a presents the block diagram of the proposed MRM^{2}DF FFT kernel, comprised of four modules (Modules 1–4). The operations associated with Modules 1–4 are illustrated in Figure 6b (association with color of Modules). Figure 6b shows that Module 1 is tasked with 4/5/6-path BF2 (radix-2) operations. Module 4 provides eight-path first and second BF8 (radix-8) computations.

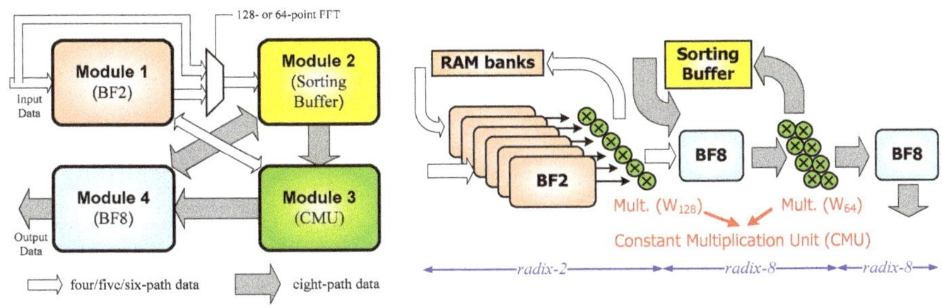

(a) (b)

Figure 6. (**a**) Block diagram of proposed MRM^{2}DF FFT kernel; (**b**) module-associated operations.

Intermediate data transfer for radix-2/8/8 calculations was achieved using a sorting buffer in Module 2. As mentioned previously, W_{128} and W_{64} twiddle factors must be multiplied with the output of the radix-2 and first radix-8 stage. Module 3 was a constant multiplication unit (CMU) for performing twiddle factor multiplications by using a novel resource-sharing scheme. Modules 1–4 are detailed in Section 3. For 64-point FFT, only Modules 2 and 3 were used to perform radix-8/8 operations.

For example, for the six-stream 1024-point MDF FFT, Figure 7 shows an operation flow to generate a 128-point FFT block based on the MDF processing (Figure 5). In most OFDM standards (e.g., IEEE 802.11ac for WiFi), the GI period is 1/4 or 1/8 of the symbol duration, depending on the channel conditions. For example, six streams of 1024 data samples were accessed to perform 1024-point FFT operations, and the number of GI samples was 256 (i.e., 1/4 symbol duration). Figure 5 shows that this mechanism allowed the calculation of six-stream 1024-point FFT through the radix-2^3 stages and 128-point FFT block. The corresponding radix-2^3 signal flow chart with the radix-2 (BF2) operations is shown in Figure 7 for reference. Through the three stages of MDF-based radix-2 operations (including twiddle factor multiplication), sets of 512/256/128-point FFT sequences were generated (corresponding with the color and slash in the signal flow chart). Finally, eight sets of six-stream sequences were generated, including 128 data and 32 GI′ samples. For each set of 128-sample data streams, the proposed MRM^{2}DF kernel was prepared to process six-stream 128-point FFT. Figure 8 presents the detailed timeliness diagram of MRM^{2}DF hardware operations and signal flow chart based on the radix-2/8/8 algorithm. The inclusion of GI duration in the time available for processing OFDM FFT is an efficient approach. However, for conventional MDC/MDF FFT structures operating at the input sample rate [20–22,25,26], the operations were generally idle during GI clock cycles, thereby reducing hardware efficiency. By contrast, the proposed MRM^{2}DF scheme used all 160 clock cycles, including GI′ (Figure 8), for 128-point FFT operations.

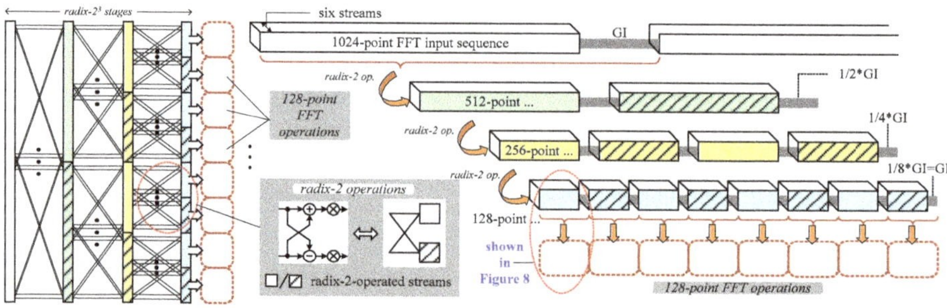

Figure 7. Operation flow for the generation of 128-point FFT block based on 1024-point MDF FFT.

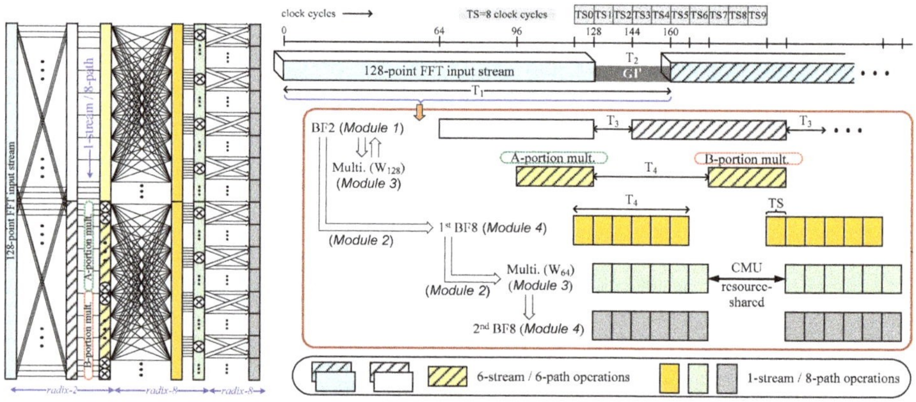

Figure 8. Detailed timeliness diagram for operations of the proposed MRM^{2}DF-structured FFT kernel and its associated signal flow chart for the radix-2/8/8 algorithm.

Figure 8 illustrates that BF2 and W_{128} multiplications were performed concurrently for all six data streams. BF8 and W_{64} multiplications were performed on eight stream-by-stream, based on a time unit of eight cycles (i.e., a time slot (TS)) for the operation of each stream. The radix-2/8/8 signal flow chart reveals that only half of the BF2 outputs (i.e., BF2 subtraction terms) were multiplied using W_{128} twiddle factors. This observation indicated that W_{128} multiplication was performed only during some of the 160 cycles (A- or B-portion mcl in green or red colors). This observation ensures that the non-occupied duration is available for W_{64} multiplication by using the resource-shared CMU. Moreover, using the TS unit (eight cycles) for BF8/W_{64} operations per stream allowed the efficient use of GI′ for the processing of additional streams. In the aforementioned example, GI′ had 32 cycles (i.e., T_2), and T_3 was half of T_2. Thus, six TS units (T_4) were available for BF8/W_{64} operations for six-stream 128-point FFT. Similar schemes can be applied to other GI′ situations by evaluating T_1–T_4 terms in Figure 8. Table 1 lists T_1–T_4 parameters associated with various operating modes of 128-point FFT. In the 64-point FFT configuration (Figure 6), the proposed scheme provided only BF8/W_{64} operations based on the BF8/W_{64}-related portions of the timing schedule (Figure 8).

Table 1. Parameters for various operating modes.

GI′ Clock Cycles	No. of Streams (max.)	T_1 Clock Cycles	T_2 Clock Cycles	T_3 Clock Cycles	T_4 Clock Cycles
32 (1/4 symbol)	6	160	32	16	48 (6 TS)
16 (1/8 symbol)	5	144	16	8	40 (5 TS)
0 (None)	4	128	0	0	32 (4 TS)

3.2. Architecture Modules

3.2.1. Module 1

The block diagram in Figure 9 shows that Module 1 comprises six sets of BF2 units and two banks of 32-element RAM modules for radix-2 operations. The path-routing control of multiplexers (M1–M5) allows Module 1 to perform six-path addition and the subtraction of the input for BF2 execution or route the read/write the data of six-set RAM banks for I/O data delivery. Module 1 operations were more complex than those conventionally used for SDF or MDF radix-2 stages [9,24–26] because this process enables data access for W_{128} multiplications by using the shared CMU (Figure 8).

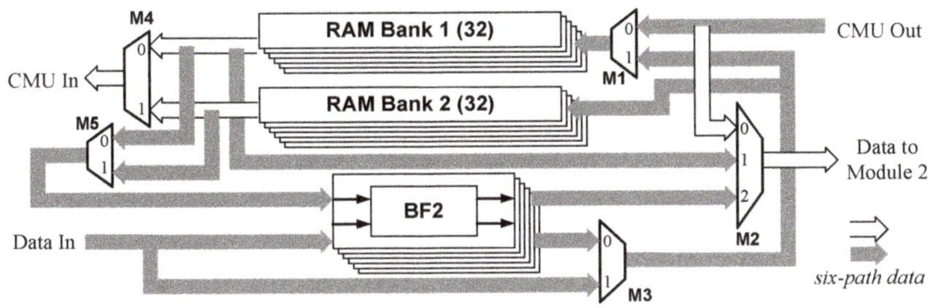

Figure 9. Block diagram of Module 1.

Referring to the top radix-2 related portions in Figure 8, Figure 10 details the operations of Module 1 and the controls of multiplexers M1–M5 for the first (the initial) and second 128-point input sequences of the six streams. Figure 10 presents addresses based on the schedule of clock cycles and radix-2 operations with W_{128} multiplication (Figure 8).

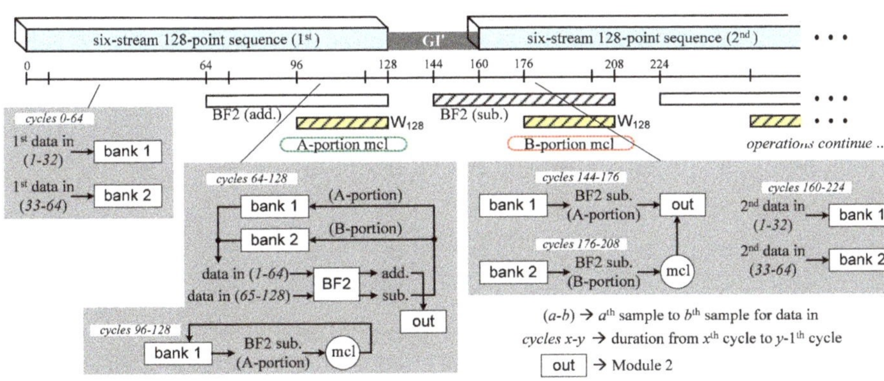

Mux \ cycles	0~32	32~64	64~96	96~128	128~144	144~160	160~176	176~208	208~224	
M1	1	X	1	0	X	X	1			
M2	X		2		X	1		0	X	
M3	1		0		X	1				• • •
M4	X		X	0	X	X		1	X	
M5	X		0	1	X	X			X: don't care	

Figure 10. Detailed operations of Module 1 and control of multiplexers (M1–M5) based on clock cycles.

3.2.2. Module 2

Figure 11 presents the block diagram of Module 2 (i.e., the sorting buffer), comprising six chains of shift registers (SRs) and multiplexers. The six streams of data from Module 1 were sent or routed to six SR chains (1–6) through paths A to F, respectively. Module 2 conducted the intermediate data sorting of raidx-2/8/8 operations. For the radix-2/8 calculation, six-stream radix-2 data from Module 1 were stored in six-chain SR and then selectively read out to Module 4 for the first BF8 operations. The calculated results were stored back into the SR chain 1 by using configurable path routing and were prepared for subsequent radix-8/8 calculation. To continue radix-8/8 operations, eight data paths were selected from SR chain 1 and the output to Module 3 (CMU) for W_{64} multiplication and the subsequent execution of the second BF8. Figure 11 shows each SR chain was structured using eight SR units, which were classified into two types: Type I and Type II. Both types exhibited eight stages of flip flops and front multiplexers to support the data shift and hold. Type I was used for SR chain 1, whereas providing parallel data accesses for eight flip flop stages. Type II was used for SR chains 2–6 and optionally enabled the routing input from other SR units.

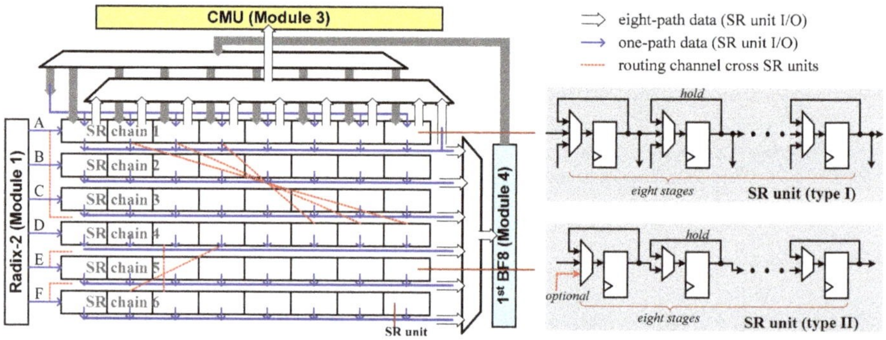

Figure 11. Block diagram of Module 2 (sorting buffer).

Figure 12 details the sorting operations of Module 2 corresponding to the periodic 10 TS units (i.e., TS0 to TS9) in Figure 8. In Figure 12, the data of each of the six A, B, … , or F streams from Module 1 (Figure 11) were represented using eight 8-sample data sections (i.e., X0, X1, … , and X7, where X can refer to A, B, … , or F). A, B, … , or F indicate the current six streams, whereas A′, B′, … , or F′ denotes the next six streams. Figure 12 at TS0 show that seven data sections of each stream (e.g., A0-A6 and B0-B6) were prepared in SR chains. From TS0 to TS5, the first BF8 operation was performed for the eight-section A, B, … , and F sequence, and the calculated results were stored back to SR chain 1. At next TSs (i.e., TS1–TS6), the first BF8 outcomes stored in SR chain 1 for B-F streams were sent out for CMU and for second BF8 operations. The data access schemes of SR chain 1 for first BF8 (stored in) and CMU/second BF8 (sent out) were alternately changed to satisfy the radix-8/8 data permutation [Equation (5)]. This observation was enabled using type I SR (Figure 11), which proved dual serial/parallel data shifts. Figure 12 shows the data of the new-coming six sequences, namely A′, B′, … , and F′, were sent to Module 2 after TS3. By appropriately routing the data across the SR chains (the red dashed line in Figures 11 and 12), the eight-section data of A′, B′, … , and F′ streams were regularly arranged in SR chains for new preparation at TS7/TS8/TS9. Therefore, the proposed sophisticated sorting buffer scheme allowed Module 2 to efficiently access intermediate data for radix-2/8/8 operations. The number of word storage elements for Module 2 was 64 per stream. By combing the level of two 32-element RAMs for Module 1, the total number of storage elements was 128. This process facilities the use of minimum memory sizes for 128-point FFT as the level of SDF/MDF-based designs [9,24–26].

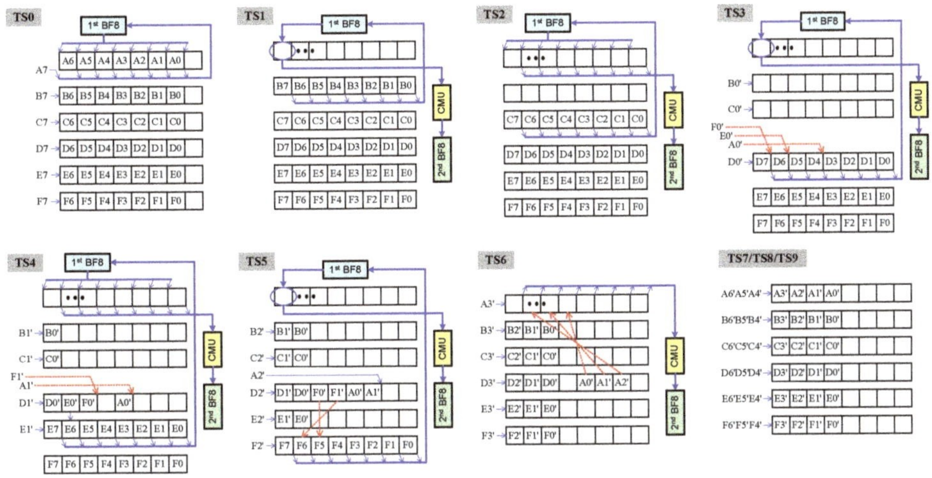

Figure 12. Details of operations in TS0-TS9 (Figure 8) for Module 2 (sorting buffer).

3.2.3. Module 3

Module 3 (CMU) performed intermediate W_{128} and W_{64} multiplications for the radix-2/8/8 algorithm. According to the $\pi/4$ symmetry of twiddle factors [32], W_{64} multiplications could be calculated using only eight factors of $W_{64}^{p'} = \exp(-j2\pi p'/64)$ and $p' = 1, 2, \ldots, 8$, in cooperation with sweeping and sign controls [24,25]. Considering the six eight-cycle TSs (i.e., TS1–TS6) executed by the second BF8 in Figure 12, the CMU/W_{64} multiplication for each cycle within a single TS period could be performed using the eight-path $W_{64}^{p'}$ factors, which is shown using p' parameters in Table 2. The appropriate rescheduling of cycles 2, 4, and 6 (the red arrow in Table 2) could be used to reduce conflict in $W_{64}^{p'}$ multiplication in order to minimize the number of required $W_{64}^{p'}$ constant multipliers. Additional multiplexer controls and registers were introduced to Module 2 to enable rescheduling. For W_{128} multiplication, $\pi/4$-symmetry check increased the required number of $W_{128}^{p'}$ factors to sixteen ($p' = 1, 2, \ldots, 16$) compared with $W_{64}^{p'}$. However, the scheme in [33] can be employed to decompose $W_{128}^{p'}$ into a form as Equations (6) and (7) if p' parameter is even or odd. Because $W_{64}^{k,k\pm1}$ could be obtained using one of the original $W_{64}^{p'}$ factors, thus, $W_{128}^{p'}$ could be derived as a combined calculation of W_{128}^{1or3} and $W_{64}^{p'}$. Therefore, only W_{128}^{1or3} constant multipliers were required to perform W_{128} multiplication with the existing $W_{64}^{p'}$ constant multipliers.

$$W_{128}^{p'} = W_{128}^{2k} = W_{64}^{k}, \, when \, p' \, is \, even \tag{6}$$

$$W_{128}^{p'} = W_{128}^{2k+1} \begin{array}{l} = W_{128}^{1} W_{64}^{k} = W_{128}^{-1} W_{64}^{k+1} \\ = W_{128}^{3} W_{64}^{k-1} = W_{128}^{-3} W_{64}^{k+1}, \, when \, p' \, is \, odd \end{array} \tag{7}$$

As mentioned previously (Figures 9 and 10), six streams of data were accessed between Modules 1 and 3 for W_{128} multiplications. To avoid conflict operations associated with access to identical six-stream W_{128} factors in the same cycle, a one-sample shift was sequentially applied to the six paths of the CMU-related data accesses of the RAM banks (from Module 1, Figure 9). Figure 13 shows that the architecture of Module 3 allows access to six streams of data from Module 1 for multiplications involving W_{128} factors, and eight data paths from Module 2 for multiplications involving W_{64} factors (two arrow lines in Figure 13). Several duplicated W_{128}^{3} and W_{64}^{4} constant multipliers were used to manage residual conflicting W_{128} and W_{64} multiplications, such as cycle 4 in Table 2, to perform W_{64}^{4} multiplication twice. On the basis of our evaluation, the overall CMU (Module 3) covered an area equivalent to 4.18 complex multipliers (structured as four real multipliers and two adders). This process was more area-efficient than the direct approach by using six/eight complex multipliers for six-stream or eight-path operations.

Table 2. Lists and scheduling of eight-path $W_{64}^{p'}$ factors (in p') for each cycle in a given TS.

→cycles	0	1	2	3	4	5	6	7
BF8 path 1	0	0	0	0	0	0	0	0
BF8 path 2	0	1	2	3	4	5	6	7
BF8 path 3	0	2	4	6	8	6	4	2
BF8 path 4	0	3	6	7	4	1	2	5
BF8 path 5	0	4	8	4	0	4	8	4
BF8 path 6	0	5	6	1	4	7	2	3
BF8 path 7	0	6	4	2	8	2	4	6
BF8 path 8	0	7	2	5	4	3	6	1

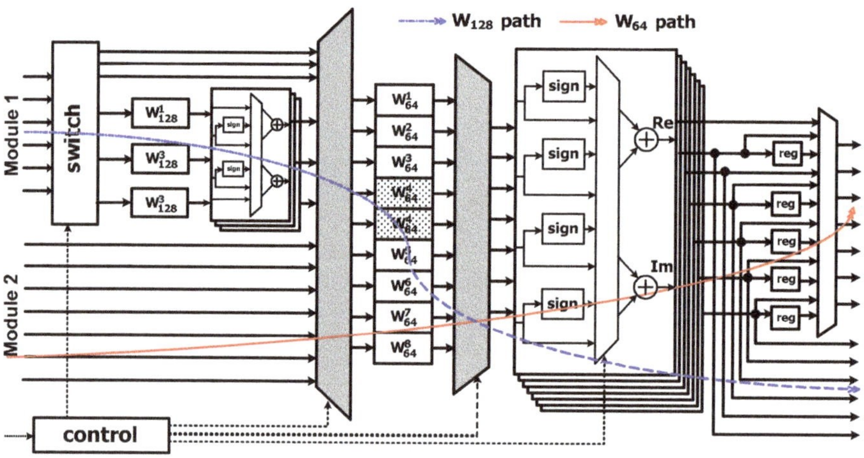

Figure 13. Block diagram of Module 3 (CMU).

3.2.4. Module 4

The two BF8 units in Module 4 were used for the first and second BF8 operations, respectively. Figure 14 shows the BF8 unit used three radix-2 stages based on the radix-2^3 algorithm [9]. In each radix-2 stage, four sets of BF2 elements were used for eight-path BF8 operations. Moreover, one W_8^1 and one W_8^3 constant multiplier each were required between the radix-2 stages 2 and 3.

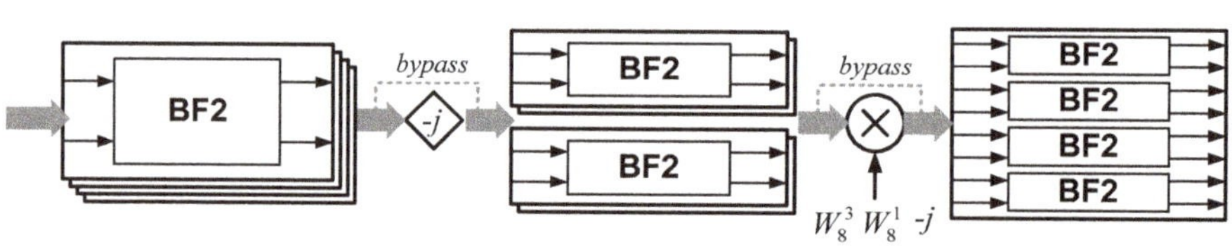

Figure 14. Block diagram of the first or second BF8 unit in Module 4.

4. Results and Comparison

4.1. Design Implementation

The evaluated signal is used to evaluate the proposed kernel accuracy. After fixed-point simulation using MATLAB, the signal-to-quantization-noise ratio (SQNR) was approximately 40 dB by using the proposed FFT kernel. This was implemented in a 1.0-V UMC 90-nm 1P9M complementary metal-oxide semiconductor process, the proposed FFT kernel used the Synopsys Design Compiler to synthesize the RTL code and employed Cadence SOC Encounter for placement and routing. The proposed FFT kernel consumed 10.72 mW of power and was operated at 80 MHz, and the core area of the proposed kernel was 739 × 734 μm^2. Figure 15 shows the core layout of the proposed kernel and its characteristics. The kernel test chip could perform 128/64-point FFT in 4–6 streams depending on the GI′ duration. A test module was included in this kernel chip to enable the serial storage of test patterns from the I/O and to send them to the kernel circuit along multiple paths. Moreover, the FFT results were sent to the test module and then sequentially read out through I/O for data evaluation. The post-layout simulation was performed based on the idea that the FFT kernel continuously processes the pattern stored in the test module.

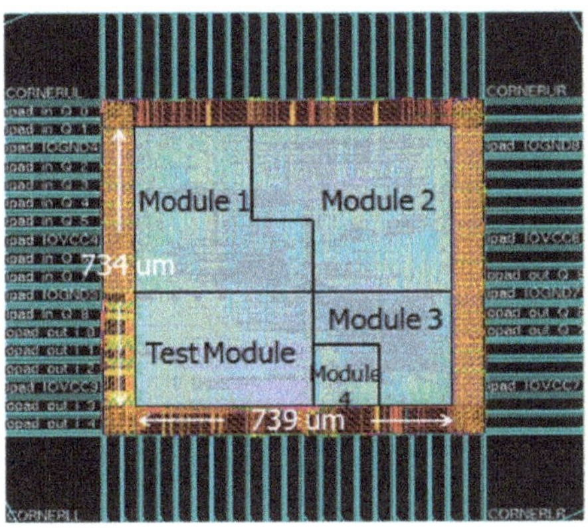

FFT modes	6-stream / 128-point
Technology	UMC 90 nm
Core Voltage	1.0 Volt
Wordlength (IO/Internal)	12/16 bits
SQNR	40.17 dB
Clock Rate	80 MHz
Core Area	0.55 mm^2
Power Consumption	10.72 mW (@ 80 MHz)

Figure 15. Layout and performance summary of FFT kernel test chip.

4.2. Comparison

Table 3 compares the proposed design with previous studies which could perform the same FFT operations (i.e., maximum FFT length of 128). An SDF 128-point FFT processor was developed based on the radix-2^4 algorithm [30] and its performance was extended to four streams for evaluation. That which is compared deals with devices based on different technologies; therefore, the area was normalized [Equation (8)] based on the scheme in [27,33]. Although the GI duration is relatively short (e.g., 1/4 symbol duration) in timeliness considerations, the proposed FFT kernel could lend support to a throughput gain of (6/4) for the number of streams, only mainly needing additional memory elements for the two-stream data storage. The comparison results demonstrated the per-stream area efficiency of the proposed design by using area-efficient Modules 2 and 3 (sorting buffer and CMU), resource sharing and hardware use GI duration. Table 3 shows the improved area efficiency of the proposed design because the proposed scheme achieved the highest throughput by using modest hardware resources (i.e., throughput per area).

$$Nor.\,Core\,Area = \frac{Core\,Area}{(Tech./90\,nm)^2} \tag{8}$$

Table 3. Performance evaluation and comparison based on 128-point FFT specification.

	[25]	[30] (ref.)	[21]	Proposed
Architecture	MRMDF	R-2^4 SDF	MDC	MRM^{2}DF
Technology	0.13 μm	90 nm	0.18 μm	90 nm
No. stream (path)	4	4	4	6
Clock rate (MHz)	40	80	75	80
Nor. core area[3] (mm^2)	0.67 [1]	0.54	0.525 [2]	0.55 [1]
Throughput (*R*: clock rate)	4*R*	4*R*	4*R*	6*R* (six streams)
Throughput per area (*R*/mm^2)	5.97	7.41	7.62	10.91

[1] A test module was included. [2] An output sorting buffer was included. [3] All the values for area were normalized using Equation (8).

5. Conclusions

In this study, an area-efficient FFT kernel was presented using the MRM^{2}DF structure for multistandard MIMO-OFDM applications. The proposed design scheme allowed 64/128-point FFT in up to six streams through the efficient use of GI duration, thereby enhancing area efficiency per stream. Novel resource-sharing and operation-rescheduling schemes were developed using an area-reduced CMU to minimize multiplication hardware costs. Finally, a sophisticated sorting buffer was proposed using the minimum memory size to further reduce area overhead. A test chip was developed using UMC 90-nm technology, and was validated through post-layout simulation. The proposed design exhibited an area efficiency superior to that of previous 128-point-based FFT designs in terms of the throughput per area.

Author Contributions: S.-N.T. conceptualized and designed the proposed methodology. He also mainly wrote the manuscript. Y.-H.C. supported the technique consultation and co-wrote the article.

Funding: This work was supported in part by the Ministry of Science and Technology of Taiwan under project MOST 107-2221-E-033-009 and 107-2221-E-182-066, and the Chang Gung Memorial Hospital, Linkou under project CMRPD2H0301, CMRPD2H0051, CMRPD2G0312, and CIRPD2F0013.

Acknowledgments: The authors would like Hsiuan-Ting Wang and Chih-Chiao Tsai for materials preparation and technical supports in IC physical designs.

Conflicts of Interest: The authors declare no conflict of interest.

References

1. Yang, H.A. Road to future broadband wireless access: MIMO-OFDM-based air interface. *IEEE Commun. Mag.* **2005**, *43*, 53–60. [CrossRef]
2. Panajotović, A.; Riera-Palou, F.; Femenias, G. Adaptive uniform channel decomposition in MU-MIMO-OFDM: Application to IEEE 802.11ac. *IEEE Trans. Wirel. Commun.* **2015**, *14*, 2896–2910. [CrossRef]
3. Ketonen, J.; Juntti, M.; Cavallaro, J.R. Performance—complexity comparison of receivers for a LTE MIMO–OFDM system. *IEEE Trans. Signal Process.* **2010**, *58*, 3360–3372. [CrossRef]
4. Dahlman, E.; Parkvall, S.; Skold, J. *4G—LTE-Advanced Pro and the Road to 5G*; Academic Press: Cambridge, MA, USA, 2016.
5. Parkvall, S.; Dahlman, E.; Furuskar, A.; Frenne, M.N.R. The new 5G radio access technology. *IEEE Commun. Stand. Mag.* **2017**, *1*, 24–30. [CrossRef]
6. Ali, A.; Hamouda, W. A multi-mode IFFT/FFT processor for IEEE 802.11ac: Design and implementation. *Wirel. Commun. Mob. Comput.* **2016**, *16*, 1713–1725. [CrossRef]
7. Yang, C.H.; Yu, T.H.; Marković, D. Power and area minimization of reconfigurable FFT processors: A 3GPP-LTE example. *IEEE J. Solid-State Circuits* **2012**, *47*, 1–12. [CrossRef]
8. Baas, B.M. A low-power, high-performance, 1024-point FFT processor. *IEEE J. Solid-State Circuits* **1999**, *34*, 380–387. [CrossRef]

9. Shousheng, H.; Torkelson, M. Designing pipeline FFT processor for OFDM (de)modulation. In Proceedings of the URSI International Symposium on Signals, System, and Electron, Pisa, Italy, 22 October 1998; pp. 257–262.
10. Jo, B.G.; Sunwoo, M.H. New continuous-flow mixed radix (CFMR) FFT using novel in-place strategy. *IEEE Trans. Circuits Syst. Part I Regul. Pap.* **2005**, *52*, 911–919. [CrossRef]
11. Chen, C.M.; Hung, C.C.; Huang, Y.H. An energy-efficient partial FFT processor for the OFDMA communication system. *IEEE Trans. Circuits Syst. II Express Briefs* **2010**, *57*, 136–140. [CrossRef]
12. Mohanty, B.K.; Meher, P.K. Area-delay-energy efficient VLSI architecture for scalable in-place computation of FFT on real data. *IEEE Trans. Circuits Syst., Part-I Regul. Pap.* **2019**, *66*, 1042–1050. [CrossRef]
13. Tang, S.N.; Jan, F.C.; Cheng, H.W.; Lin, C.K.; Wu, G.Z. Multimode memory-based FFT processor for wireless display FD-OCT medical systems. *IEEE Trans. Circuits Syst. I Regul. Pap.* **2014**, *61*, 3394–3406. [CrossRef]
14. Xing, Q.J.; Ma, Z.G.; Xu, Y.K. A novel conflict-free parallel memory access scheme for FFT processors. *IEEE Trans. Circuits Syst. II Express Briefs* **2017**, *64*, 1347–1351. [CrossRef]
15. Liu, S.; Liu, D. A high-flexible low-latency memory-based FFT processor for 4G, WLAN, and future 5G. *IEEE Trans. Very Large Scale Integr. (VLSI) Syst.* **2018**, *27*, 511–523. [CrossRef]
16. Hwang, Y.T.; Chen, Y.J.; Chen, W.D. Scalable FFT Kernel designs for MIMO OFDM based communication systems. In Proceedings of the IEEE Conferrence TENCON, Taipei, Taiwan, 30 October–2 November 2007; pp. 1–4.
17. Hung, C.L.; Long, S.S.; Shiue, M.T. A low power and variable-length FFT processor design for flexible MIMO OFDM systems. In Proceedings of the IEEE International Symposium on Circuits and System (ISCAS), Taipei, Taiwan, 24–27 May 2009; pp. 705–708.
18. Chen, S.G.; Huang, S.J.; Garrido, M.; Jou, S.J. Continuous-flow parallel bit-reversal circuit for MDF and MDC FFT architectures. *IEEE Trans. Circuits Syst. I Reg. Pap.* **2014**, *61*, 2869–2877. [CrossRef]
19. Lin, Y.T.; Tsai, P.Y.; Chiueh, T.D. Low-power variable length fast Fourier transform processor. *Proc. IEE Comput. Dig. Tech.* **2005**, *152*, 499–506. [CrossRef]
20. Yang, K.J.; Tsai, S.H.; Chuang Gene, C.H. MDC FFT/IFFT processor with variable length for MIMO-OFDM systems. *IEEE Trans. Very Large Scale Integr. (VLSI) Syst.* **2013**, *21*, 720–731. [CrossRef]
21. Fu, B.; Ampadu, P. An area efficient FFT/IFFT processor for MIMO OFDM WLAN 802.11n. *J. Signal Process. Syst.* **2009**, *56*, 59–68. [CrossRef]
22. Locharla, G.R.; Mahapatra, K.K.; Ari, S. Variable length mixed radix MDC FFT/IFFT processor for MIMO-OFDM application. *IET Comput. Digit. Tech.* **2018**, *12*, 9–19. [CrossRef]
23. Yoshizawa, S.; Orikasa, A.; Miyanaga, Y. An area and power efficient pipeline FFT processor for 8×8 MIMO-OFDM systems. In Proceedings of the IEEE International Symposium on Circuits and System (ISCAS), Rio de Janeiro, Brazil, 15–19 May 2011; pp. 2705–2708.
24. Lin, Y.W.; Liu, H.Y.; Lee, C.Y. A 1-GS/s FFT/IFFT processor for UWB applications. *IEEE J. Solid-State Circuits* **2005**, *40*, 1726–1735. [CrossRef]
25. Lin, Y.W.; Lee, C.Y. Design of an FFT/IFFT processor for MIMO OFDM systems. *IEEE Trans. Circuits Syst. I Reg. Pap.* **2007**, *54*, 807–815. [CrossRef]
26. Liu, H.; Lee, H. A high performance four-parallel 128/64-point radix-2^4 FFT/IFFT processor for MIMO-OFDM systems. In Proceedings of the IEEE Asia Pacific Conference on Circuits and System (APCCAS), Macao, China, 30 November–3 December 2008; pp. 834–837.
27. Chen, Y.; Lin, Y.W.; Taso, Y.C.; Lee, C.Y. A 2.4-Gsample/s DVFS FFT processor for MIMO OFDM communication systems. *IEEE J. Solid-State Circuits* **2008**, *43*, 1260–1273. [CrossRef]
28. Tang, S.N.; Liao, C.H.; Chang, T.Y. An area- and energy-efficient multimode FFT processor for WPAN/WLAN/WMAN systems. *IEEE J. Solid-State Circuits* **2012**, *47*, 1419–1437. [CrossRef]
29. Liu, L.; Ren, J.; Wang, X.; Ye, F. Design of low-power, 1GS/s throughput FFT processor for MIMO-OFDM UWB communication system. In Proceedings of the IEEE International Symposium Circuits and System (ISCAS), New Orleans, LA, USA, 27–30 May 2007; 2007; pp. 2594–2597.
30. Oh, J.Y.; Lim, M.S. Area and power efficient pipeline FFT algorithm. In Proceedings of the IEEE Workshop on Signal Processing Systems (SiPS), Athens, Greece, 2–4 November 2005; pp. 520–525.
31. Mahdavi, M.; Edfors, O.; Öwall, V.; Liu, L. A low latency FFT/IFFT architecture for massive MIMO systems utilizing OFDM guard bands. *IEEE Trans. Circuits Syst. I Reg. Pap.* **2019**, *66*, 2763–2774 [CrossRef]

32. Maharatna, K.; Grass, E.; Jagdhold, U. A 64-point Fourier transform hip for high-speed wireless LAN application using OFDM. *IEEE J. Solid-State Circuits* **2004**, *39*, 484–493. [CrossRef]
33. Tang, S.N.; Tsai, J.W.; Chang, T.Y. A 2.4 GS/s FFT processor for OFDM based WPAN applications. *IEEE Trans. Circuits Syst. II Express Briefs* **2010**, *57*, 451–455. [CrossRef]

applied sciences

MDPI

Article

The Design and Building of a Hexapod Robot with Biomimetic Legs

Min-Chan Hwang *, Feifei Liu, Jie Yang and Yuanzhang Lin

School of Electrical Engineering and Automation, Jiangxi University of Science and Technology, Ganzhou 341000, China
* Correspondence: aemchwang@163.com; Tel.: +86-1777-075-4280

Received: 1 June 2019; Accepted: 9 July 2019; Published: 11 July 2019

Abstract: A hexapod robot with biomimetic legs was built to implement a distributed control system, where a mechanism is proposed to serve as the central pattern generator and a computer to act as the brain-stem, cooperating with the central pattern generator through wireless communication. The proposed mechanism is composed of two modules, i.e., the tripod gait generator and the Theo Jansen Linkage. The tripod gait generator is a device that uses a single motor to generate a tripod gait, while the Theo Jansen Linkage rhythmically executes the legged motion. In a sense, we are trying to implement the locomotion of a robot by means of a hybrid computational system, including the mechanism part and the electronic processors part. The complex mathematical function of the foot movement is realized by the ensemble of links of the Theo Jansen Linkage, so as to alleviate the computational burden. Besides, the proposed design, based on non-collocated actuators, is intended to minimize the number of actuators while reducing the building cost of the robot.

Keywords: central pattern generator; Theo Jansen Linkage; non-collocated actuators

1. Introduction

The grounds of construction sites or factories are often muddy or scattered with debris. Under such circumstances, wheeled vehicles used for material handling might be easily blockaded by obstacles or experience skidding on the muddy grounds. Hence, this inspires us to pursue the development of a hexapod robot to replace a wheeled vehicle. However, robots are complex and expensive machines, consisting of many actuators, sensors, transmissions, and hardware. Therefore, a method of developing a hexapod robot to reduce the cost by means of using a minimal number of actuators is proposed. In this article, a hexapod robot was conceived, designed, and built.

In the development of a legged robot, there are two primary concerns, i.e., how to generate a stable gait, so that the robot can walk without tumbling, and how to perpetually generate stable gaits. Regarding the first concern, McGhee and Frank [1,2] proposed the COG (center of gravity projection) method in 1968, stating that the legged robot is statically stable if the horizontal projection of its COG lies within the support polygon, which is defined as the convex polygon formed by connecting the footprints. Orin [3] generalized the COG method in 1976, proposing the COP (center of pressure) method, where a robot is dynamically stable if the projection of the COG, along the direction of the resultant force acting on the COG, lies within the support polygon. In 1969, Vukabratovic and Juricic [4,5] further proposed a method in favor of the biped robot called the ZMP (zero moment point) method, where a robot is stable if the moment about the COP, at its supporting foot, is zero. As to the second concern, many rhythmic movements such as locomotion, respiration, swallowing, etc., in animals, have been found to be produced by a CPG (central pattern generator) [6–11].

Neuroscientists [12] have employed a variety of techniques, including anatomical, behavioral, physiological methods, etc., to investigate the specific neural circuits and discover the mapping function in those circuits. One notable study on the locomotion of the salamander [13] proposed the CPG model based on nonlinear oscillators instead of neural network oscillators, presented numerical as well as mechanical simulations, and successfully constructed a salamander-like robot. Some studies [14–16] have addressed to the locomotion of hexapods with CPG and established mathematical models not only including the rhythmic generator, but also the interlimb coordinator, so that the hexapod can adapt to variant terrains by gait transition. Beyond bio-inspiration, a new trend of research that is noteworthy is the merging of natural and artificial components, which is defined as bio-hybrid organisms [17,18], also called bio-robots, in which an artificial component or a biological organ is incorporated into an animal or robot, respectively. These studies even investigate how the artificial agent interacts with an animal individual or a population. So far, all studies have relied on electronic circuits or processors to implement the locomotion of legged robots.

Computers have always been thought of as nothing more than electronic devices, which is not necessarily the case. Recalling the evolution of computer science, the earlier computing devices, called calculators, invented by Blaise Pascal of France (1642) and G.W. Leibniz of Germany (1671), were built with the technology of gears. In fact, they were mechanical calculators, where data was represented through gear positioning and entered mechanically, by adjusting the initial gear positions. The output of the calculators was achieved by observing the final gear positions. Therefore, mechanical devices can be regarded as computational processors, as long as they execute certain mathematical operations. This was the inspiration to create a mechanism, or mechanical computational processor, to complement the electronic processors.

Moreover, neuroscience discovered that the CPG, located in the spinal cord, is an autonomous device, almost requiring neither the peripheral sensor feedback, nor the regulation command from the brain-stem. Therefore, a hexapod robot with biomimetic legs was built, to implement such a distributed control system, where a mechanism is proposed to serve as the CPG and a computer acts as the brain-stem, to command the autonomous device through wireless communication. In a sense, we are trying to implement the locomotion of a robot by means of a hybrid computational system, including a mechanism and electronic computers. The proposed mechanism comprises two modules, i.e., the tripod gait generator and the Theo Jansen Linkage. The tripod gait generator is a device that uses a single motor to generate a tripod gait that couples the middle leg on one side with the front and rear leg on the other side, while the TJL (Theo Jansen Linkage) rhythmically executes the legged motion.

The TJL was first introduced by Theo Jansen in 1990, where he presented a strandbeest [19], elegantly achieving a bio-inspired locomotion. This soon drew the attention of robotics researchers [20–25]. The TJL adopted in this paper is an eight-bar linkage. The interesting point is that the foot movement is a complex mathematical function, from a crank to a rocker. Nonetheless, it can be simply realized by the combination of these eight links. Consider that, if the same function is processed by an electronic device, it will consume a great amount of computational time and memory resources. Therefore, our proposed mechanism can alleviate the computational burden. Moreover, most hexapods are designed with collocated actuators, i.e., each joint is mounted with an actuator, so the number of actuators is usually high, reaching even 18, in number. Using a high number of actuators leads to many adverse effects, including increased challenge for the algorithms to control legged motions, degradation of the loading capacity, and increase of the construction cost. Hence, the present proposed design is based on non-collocated actuators, so as to minimize the number of actuators while reducing the building cost of the robot.

2. Mechanical Structure

The TJL can be built using either 8 or 12 links, depending on the fabrication method of two triangular links of the TJL. If each triangular link is machined into a whole piece, the TLJ is an 8-bar linkage, whereas if each of them is assembled by three straight bars, the TLJ is a 12-bar linkage. The TJL, as adopted in this paper, is the eight-bar linkage option. It can be considered as a device to implement the mathematical function for the legged motion, as will be presented in the Kinematics section. As illustrated in Figure 1, the complexity of the foot movement implementation is not in any way reduced, but rather a passing from the computer to the mechanism occurs, i.e., coding is replaced by component designing. Once the ensemble of links is determined, it can generate a deterministic orbit.

(a) (b)

Figure 1. (**a**) Assembly drawing. (**b**) Mechanical components.

The hexapod robot (Figure 2a) is divided into three modules, i.e., an upper deck, a bottom deck, and a swivel connecting the both of them (Figure 2b). In regard to manufacturing or maintenance, we should avoid diversifying the components design-wise. Instead, it is more practical to design the components shared by different modules. Hence, both decks are designed according to the same structure except their legs, which are mounted with opposing orientation. Therefore, it is adequate to study just one of them, while the bottom deck will be used for illustration purposes.

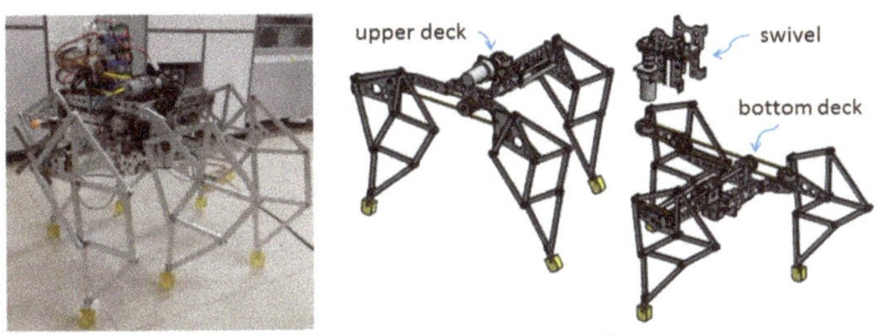

(a) (b)

Figure 2. (**a**) Completed hexapod robot. (**b**) Three major modules.

The bottom deck consists of one tripod gait generator and three TJLs. The tripod gait generator is a module dispatching power from a motor to three legs (Figure 3a), each of which is a TJL.

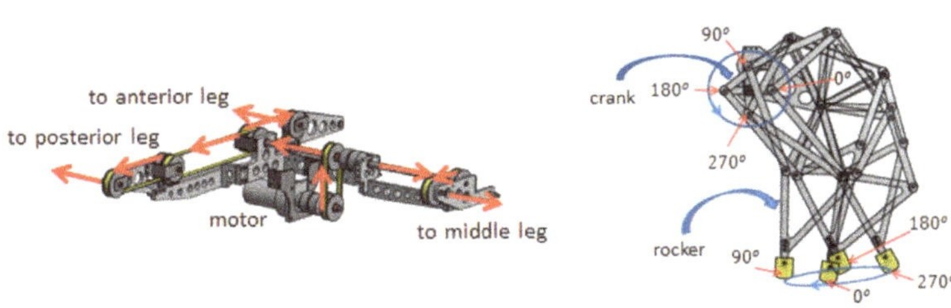

(a) (b)

Figure 3. (**a**) The module of a tripod gait generator. (**b**) Theo Jansen leg with four phases.

Let the crank of a TJL pose at four different angles, i.e. 0°, 90°, 180° and 270°. Consequently, these four postures of the TJL are superimposed, showing the motion of the rocker in relation to the crank. In Figure 3b, the tip of the rocker draws an orbit, in a sense opposite to the rotation of the crank, as the four phase angles of the crank, labeled at the foot trajectory, help acknowledge how the rocker is related to the crank during motion.

The tripod gait module is also a device which couples the middle leg on one side with the front and rear on the other side, to rhythmically generate tripod gaits. The locomotion of the hexapod is achieved by alternating two support polygons, each being the triangle connecting the tips of the rockers (Figure 4).

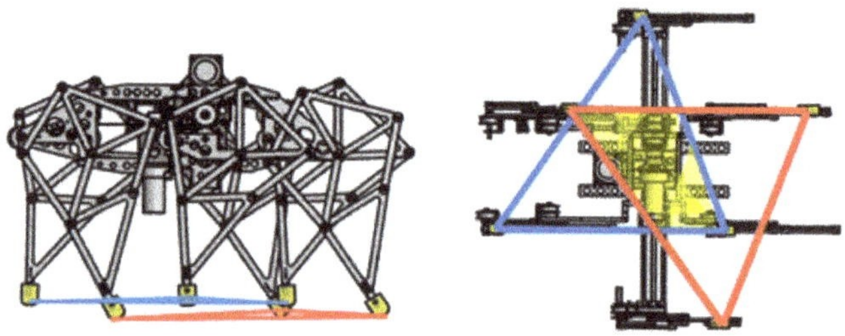

(a) (b)

Figure 4. (**a**) Lateral view of two support triangles. (**b**) Bottom view of two support triangles.

3. Kinematics

The kinematics of the TJL (Figure 5) can be derived according to the Denavit–Hartenberg [26] convention, along with certain trigonometric constraints. Hence, the orbit of the foot tip is a function of the input variable θ, also referred to as the crank angle, and defined as follows:

$$x(\theta) = A_1 cos\theta_1 + L_4 \cos(\theta_1 + \theta_2) + S_2 \cos(\theta_1 + \theta_2 + \theta_3) \quad (1)$$

$$y(\theta) = -[A_1 sin\theta_1 + L_4 \sin(\theta_1 + \theta_2) + S_2 \sin(\theta_1 + \theta_2 + \theta_3)] \quad (2)$$

where

$$S_0 = \sqrt{S_x^2 + S_y^2} \quad (3)$$

$$\beta_1 = \tan^{-1}(\frac{S_y}{S_x}) \quad (4)$$

$$\beta_2 = \beta_1 + \theta \quad (5)$$

$$S_1 = \sqrt{S_0^2 + L_1^2 - 2L_1 S_0 \cos\beta_2} \quad (6)$$

$$\beta_3 = \tan^{-1}(\frac{L_1 \sin\beta_2}{S_0 - L_1 \cos\beta_2}) \quad (7)$$

$$\beta_4 = \cos^{-1}(\frac{S_1^2 + A_2^2 - L_2^2}{2S_1 A_2}) \quad (8)$$

$$\beta_5 = cos^{-1}(\frac{A_1^2 + A_2^2 - A_3^2}{2A_1 A_2}) \quad (9)$$

$$\theta_1 = \beta_1 + \beta_3 + \beta_4 + \beta_5 - 180° \quad (10)$$

$$\beta_6 = \cos^{-1}(\frac{S_1^2 + L_4^2 - L_3^2}{2S_1 L_4}) \quad (11)$$

$$\theta_2 = \beta_7 = 360^o - (\beta_4 + \beta_5 + \beta_6) \quad (12)$$

$$\beta_9 = cos^{-1}(\frac{B_1^2 + B_3^2 - B_2^2}{2B_1B_3}) \quad (13)$$

$$\beta_{10} = tan^{-1}(\frac{(B_3 + E)sin\beta_9}{B_1 - (B_3 + E)cos\beta_9}) \quad (14)$$

$$\theta_3 = 180^\circ - (\beta_7 + \beta_{10}) \quad (15)$$

$$S_2 = \sqrt{(B_3 + E)^2 + B_1^2 - 2B_1(B_3 + E)cos\beta_9}\,. \quad (16)$$

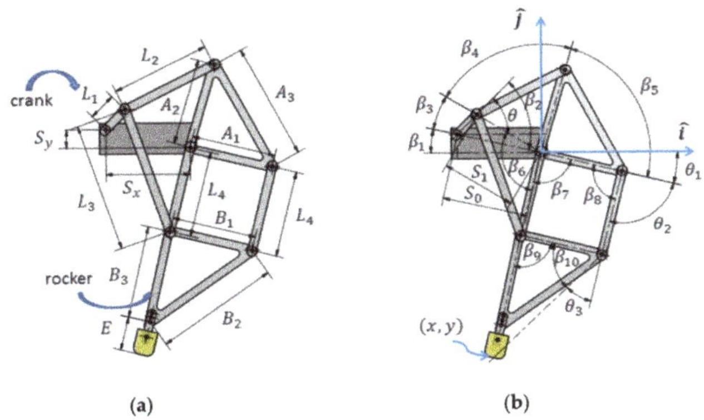

Figure 5. Theo Jansen Linkage (**a**) dimensions and (**b**) geometrical variables.

The orbits patterns produced by TJLs might end up as bell curves, ovals, sharp-pointed ovals, or lemniscates, depending on the assembly of their variant sizes. Not all generated orbits are suitable to serve as foot trajectories for the legged robots. Hence, dimensioning the TJL appropriately is nontrivial, since the orbits patterns generated by a TJL may vary according to its assembled dimensions, and can be cast into four groups, including bell curves, ovals, sharp-pointed ovals, and lemniscates. In general, the ovals or bell orbits are legitimate, while the sharp-pointed ovals are partly legitimate, and the lemniscates are illegitimate. After analyzing and balancing requirements, the design data of L_4, A_1, A_2, A_3, B_1, B_2, B_3, E, S_x, and S_y were set to be 140, 135, 140, 190, 135, 200, 150, 55, 135, and 30, respectively.

In Table 1, four cases of links with different lengths: L_1, L_2, and L_3, are illustrated along with the respective data of foot trajectories, demonstrating how they are influenced by the variant links. Although all four cases include legitimate orbits, there are some other concerns to be taken into account in order to achieve a satisfactory design. The maximum step size is defined as the distance between the leftmost point and the rightmost point. Similarly, the maximum step height is defined as the distance between the topmost point and the bottommost point. Although both of these are different from the actual step size and height, they are able to show the effectiveness of the following actions. Regarding the actual step size and height, a slightly more complicated calculation is required, which will be presented in the next section.

Indicatively, in the trajectory of Case 1 (Figure 6a), where the inclination is too deep, the corrective measure, taken by Case 2, is to reduce the lengths of L_2 and L_3. Nonetheless, as mentioned earlier, the stability of the hexapod is ensured by the intersection of two support polygons. Given the fact that, the larger the stride, the smaller the intersected polygon will be, the hexapod will eventually be destabilized if the stride is too large. In order to ensure stability, the action taken by Case 3 is to shorten the step size, by reducing the length of L_1 (Figure 6b). After these efforts, the hexapod is ready to walk. However, its step height is relatively small, so the further action, taken by Case 4, is to increase the step height (Figure 6c), by the fourth set of data in Table 1.

Table 1. Four sets of design data for L_1, L_2, and L_3.

Case	L_1 (mm)	L_2 (mm)	L_3 (mm)	Max. Step Size (mm)	Max. Step Height (mm)
1	45	160	210	223.6	26.2
2	45	172	200	223.8	18.1
3	30	172	200	148.8	15.4
4	30	150	164	139.8	37.8

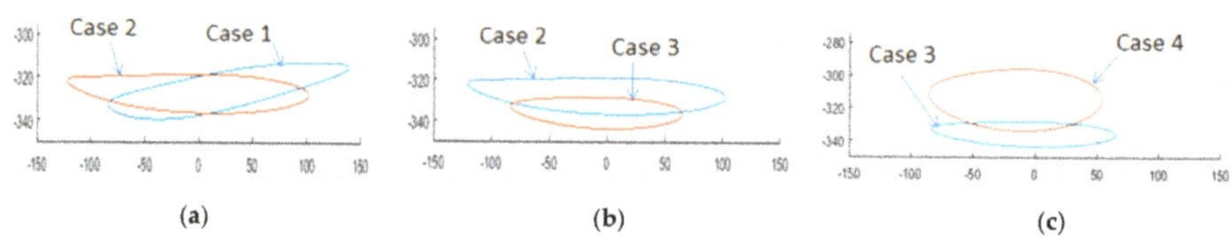

Figure 6. Actions: (**a**) inclination decreased, (**b**) step size reduced, and (**c**) step height increased.

4. Locomotion

The locomotion of this hexapod can be analyzed in terms a simplified model (Figure 7), a biped robot with absolute stability, by regarding the left and right legs of the biped version as the legs of the upper and lower deck of the hexapod, respectively. Following this, the locomotion of the robot can be characterized by the motion of its COG (Center of Gravity).

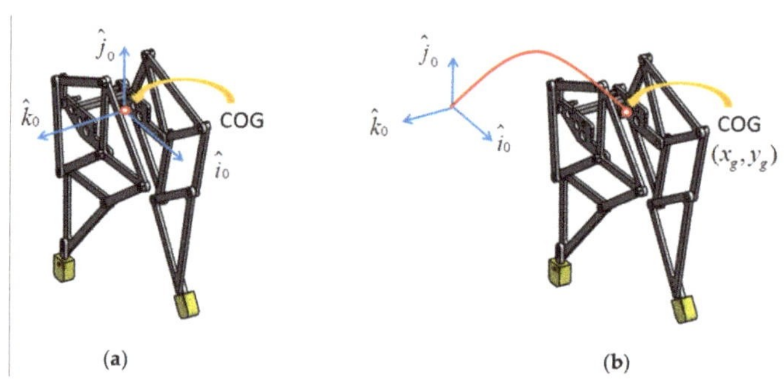

Figure 7. A simplified model of the robot: (**a**) Initial configuration and (**b**) motion of the center of gravity Projection (COG).

The cranks of the left and right legs will keep apart by 180° during rotation, and Figure 8 illustrates how the motion of the COG relates to the foot trajectory where (x,y), (x_g, y_g), L_s, H_s, and M_g stand for the foot trajectory, the motion of the COG, the step size, the step height, and the fluctuated amplitude, respectively.

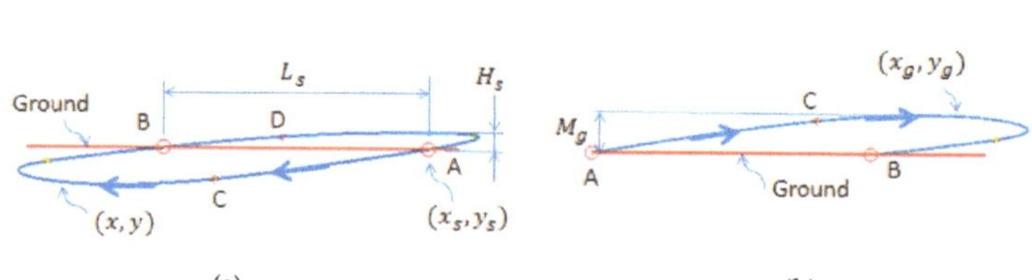

Figure 8. (**a**) Foot trajectory. (**b**) Motion of COG.

The robot is initially at rest with one foot standing at point A, while another foot stands at point B. The coordinates of point A are denoted as (x_s, y_s), which is associated with the crank angle θ_s, while the crank angle of point B is 180° apart from θ_s, i.e., $\theta_s + \pi$. Hence, point A can be obtained by determining the crank angle θ_s, where the height of point A is equal to the height of point B, using Equations (1)–(16). Following, the step size L_s, step height H_s, and fluctuated amplitude M_g, are calculated and listed in Table 2.

Table 2. Characteristics of foot trajectories related to four sets of design data.

Case	θ_s (°)	L_s (mm)	H_s (mm)	M_g (mm)	H_s/M_g
1	317.4	131.3	8.0	18.2	0.44
2	234.3	201.4	3.8	14.3	0.27
3	241.8	141.8	4.6	10.8	0.43
4	246.9	135.4	12.3	25.5	0.48

As the foot engages on point A, it pushes the robot forward, as well as upward. Hence, by mirroring the part of foot trajectory ABC, with respect to the horizontal and vertical axes, the locus of the COG is derived, yielding the following equations:

$$x_g(\theta) = nL_s + x_s - x(\theta - n\pi) \tag{17}$$

$$y_g(\theta) = y_s - y(\theta - n\pi) \tag{18}$$

$$n = \left[\frac{\theta - \theta_s}{\pi}\right] \tag{19}$$

where, the bracket [ ] yields the maximum integer, no greater than the value it contains.

The locomotion of the robot, i.e., rhythmically alternating one foot with another, can be well characterized by the locus of its COG, using Equations (17)–(19). The loci of the COG pertaining to four different cases are as shown in Figures 9–12.

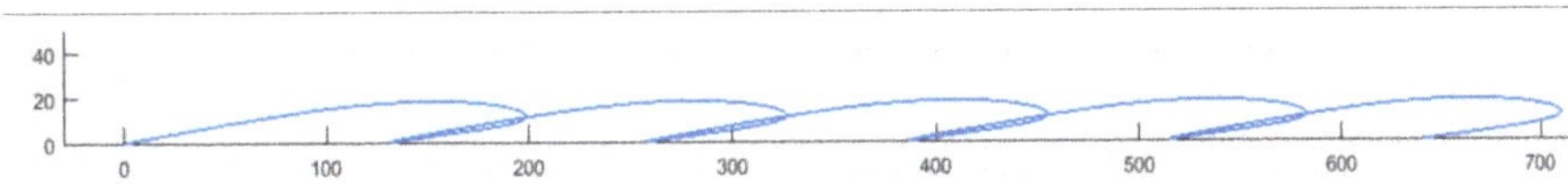

Figure 9. Locus of center of gravity for Case 1.

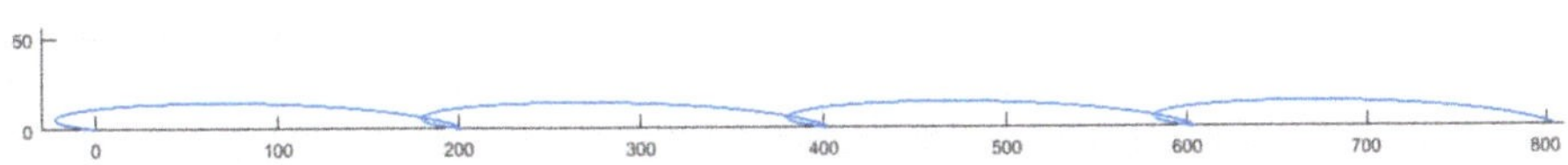

Figure 10. Locus of center of gravity for Case 2.

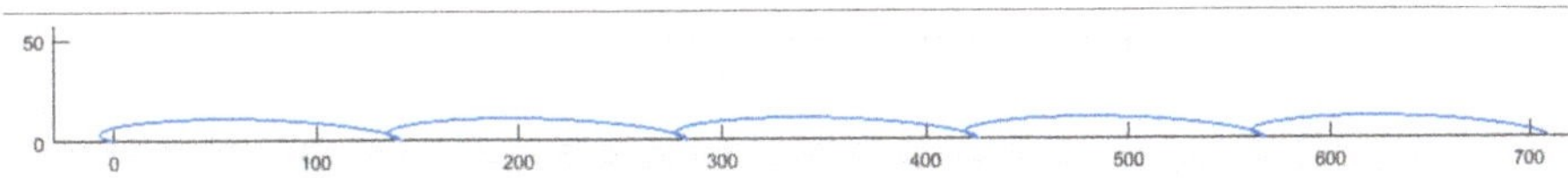

Figure 11. Locus of center of gravity for Case 3.

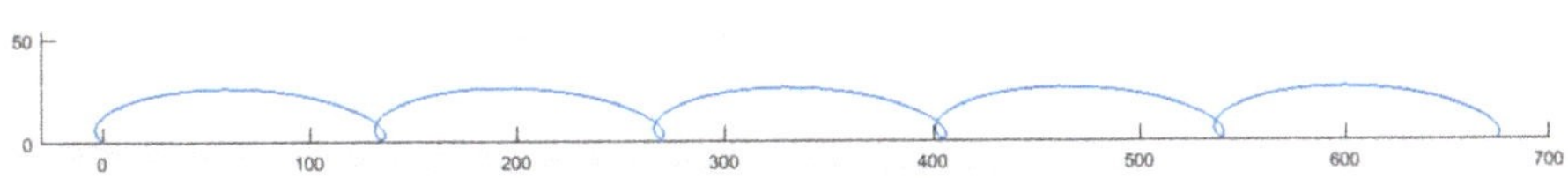

Figure 12. Locus of center of gravity for Case 4.

In Figure 9, the locus manifests that the robot will wobble significantly, while moving. This adverse effect arises from the fact that the foot trajectory, generated by the TJL in Case 1, exhibits large obliqueness. In an effort to improve this situation, the inclination of the orbit was reduced, causing a side effect on the actual step size, i.e., an increase by 1.5 times (Figures 9 and 10). The maximum step sizes in Cases 1 and 2 (Table 1) are almost the same and the respective orbits (Figure 6a) are similar. Evidently, the actual step size is not only determined by the orbit itself, but also by the points where the feet meet the ground, i.e., point A and point B (Figure 8).

Moreover, the robot can change its azimuth by a swivel mechanism, located between the upper and bottom deck. In order to avoid mechanical collision, the swivel motor is limited to rotate within ±15°. However, the robot is still able to manage turns greater than ±15° through a sequence of detailed maneuvers of its swivel and legs. An example is illustrated in Figure 13. Considering that the body and output shaft of the swivel motor are connected to the bottom and upper deck, respectively, the swivel motor has to reverse its output shaft, depending on which set of legs is standing on the ground, as the robot changes its azimuth. In Figure 13b, the swivel motor turns clockwise, to swing the upper deck in the same direction, while the legs of the bottom deck stand statically on the ground. In Figure 13c, the swivel motor turns counterclockwise to swing the bottom deck in the opposite direction, i.e., clockwise, when the legs of the upper deck stand statically on the ground.

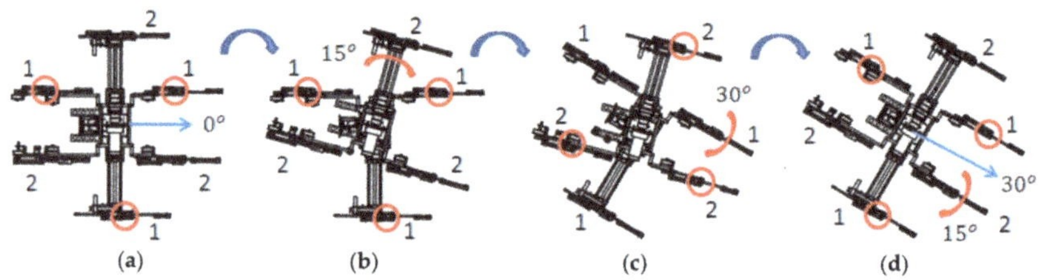

Figure 13. The robot changes its azimuth through a series of motions where the legs of the bottom deck and the upper deck are labeled as 1 and 2, respectively, while the red circles stand for the legs standing statically on the ground. (**a**) Initial posture. (**b**) The upper deck suspends its legs and makes a swing 15° clockwise, while the legs of the bottom deck stay on the ground. (**c**) The bottom deck suspends its legs and makes a swing 30° clockwise, while the legs of the upper deck stay on the ground. (**d**) The upper deck suspends its legs and makes a swing 15° clockwise, while the legs of the bottom deck stand on the ground.

5. Mechatronics of the Control System

Following the analysis and construction of the mechanism, the mechatronics of the control system (Figure 14), consisting of a microcontroller, drivers, sensors, and motors with gear trains, was built.

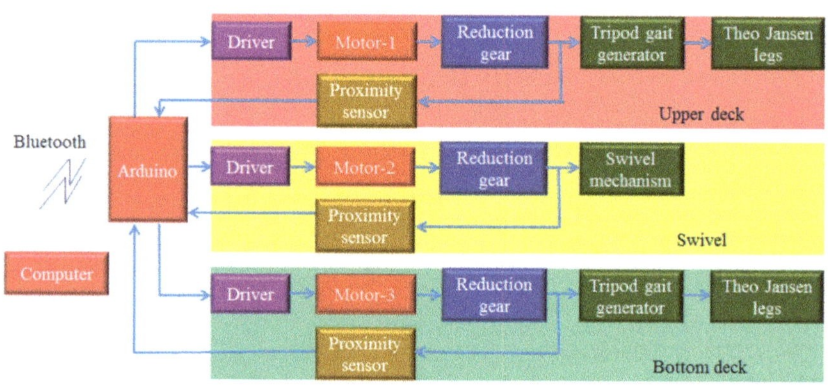

Figure 14. Block diagram of mechatronic system.

This part mainly controls the orientation of the swivel and the motions of the TJLs, with a phase of 180° between the two sets of cranks. The Arduino was chosen as the microcontroller for this project. Since the first Arduino [27] was introduced in 2005, it has become one of the most popular microcontrollers used by engineers and even large corporations. The appealing features of Arduino are mostly its openness and ease-of-use. It is worth mentioning that the programming of an embedded system is done according to cross-development principles, so that the code can run on a target boar, different from the one where the code was built. The Arduino provides an integrated development environment (IDE) for users to edit, debug, and compile their application programs on a personal computer and then upload them to the Arduino board for execution.

In Figure 15, the crank of the TJL is attached through four metal blades, evenly partitioning the circumference into four quarters. The proximity sensor, CHIIB LJ18A3-10-Z/BY, is used to detect four phases of the crank, based on which the Arduino is able to keep the two sets of cranks apart, at 180°, during operation. Regarding the control of the swivel (Figure 16), specific care needs to be taken to avoid mechanical collision. First, one system, consisting of a proximity sensor, CHIIB LJ12A3-4-Z/BY, and three metal blades, is installed to receive feedback on the azimuth of the swivel. Second, another system, using two limited switches as a safety guard to assure that the range of operation is maintained, is installed. When the swivel reaches its maximum range, it will push the leverage of the limited switch, so as to cut off the power of the swivel motor immediately. Without the safety guard system, the catastrophe of a mechanical collision is inevitable in developing the application programs. Regarding the DC motors, there are two motors with specifications of 12 V, 2 A, 35 W, 30 rpm, and 25.47 kg-cm, used for the TJLs and one motor with specifications of 12 V, 2 A, 35 W, 10 rpm, and 76.4 kg-cm, used for the swivel. The drivers are all in typical H-bridge with specifications of 12–36 V, 10 A, and 120 W.

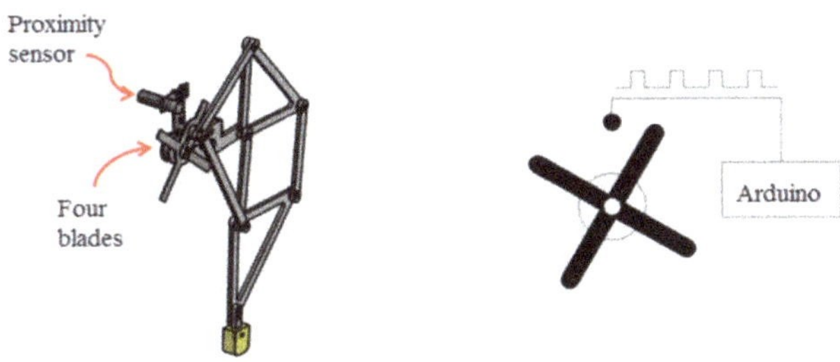

Figure 15. Sensor system to detect the phases of the crank.

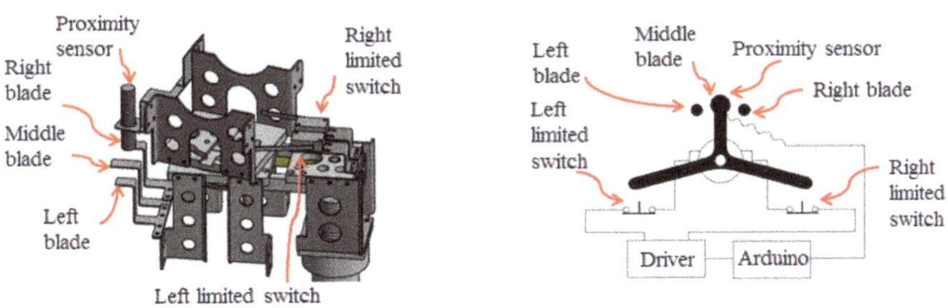

Figure 16. Sensor system to detect the azimuth of the swivel and the safety guard system.

In this project, Bluetooth (BT) was used to establish wireless communication between the computer and the Arduino. Bluetooth is a wireless technology standard, IEEE 802.15.1, for exchanging data between fixed and mobile devices over short distances, most often within 10 m, using short-wavelength

UHF (Ultra High Frequency) radio waves, in the industrial, scientific and medical radio bands, from 2.400 to 2.485 GHz, and in personal networks.

AT (Attention), commands are instructions used to control a modem. Every command line starts with "AT" or "at," which is why modem commands are called AT commands. Some AT commands used to configure the BTs and a portion of Arduino program, related to the wireless communication, are provided in Table 3. The BTs need to be set up before communication. First, the KEY-pin is set to high, before powering up the BT. As the power is turned on, the BT enters a setting mode, where its LED blinks with a period of four seconds, i.e., light on for two seconds and light off for two seconds. Then, AT commands are used to configure the baud rate, parity, stop bit, password, master/slave, binding address, etc. Following this, the power is turned off and the KEY-pin is set to low. As the power is turned on again, the BTs enter a standby mode, where they blink every second, in an attempt to pair up. At successful pairing up, both BTs get into the communication mode, where they blink twice for two seconds and light off for two seconds. Upon completion of the mechatronic control system, programs were developed, where the Arduino side mainly awaits the commands from the computer and keeps two sets of cranks apart at 180° during operation, while the computer side sends wireless commands to the Arduino. Since the legged motion has been realized by the mechanism, there is no need to develop a sophisticated algorithm in Arduino.

Table 3. AT Commands and Arduino program related to communication.

AT Commands	Arduino Program Paradigm
AT+ADDR?	header: #include <SoftwareSerial.h>
AT+PSWD?	setup: SoftwareSerial BTSerial(8, 7); //RX, TX
AT+ROLE?	Serial.begin(38400);
AT+CMODE=0/1	BTSerial.begin(38400);
AT+UART=38400,0,0	Loop: if(BTSerial.available()){
AT+BIND=98D3,31,206801	ch=BTSerial.read();
AT+BIND?	...
...	Serial.write(ch);}

In Figure 17, a sequence of snapshots of the hexapod robot, as tested on site, is illustrated. The appealing feature of this robot is that it is reliable and robust. Even if it runs out of battery or an electronics failure occurs, it can still stand because the tripod gaits are generated by the mechanism, instead of the computer.

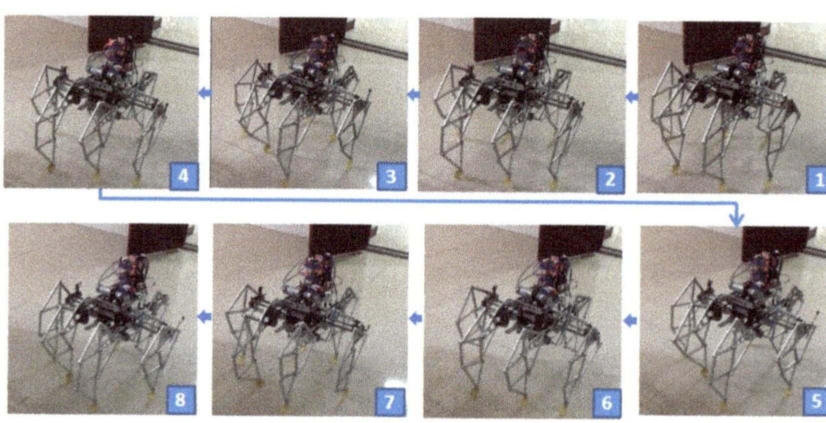

Figure 17. Snapshots of the hexapod robot in motion.

6. Comparison to Existing Hexapods

For the sake of comparison, an overview of existing hexapod robots is conducted in this section. Since the first hexapod robot with electric drivers and a computer-controller [28] was built in 1972, a variety of hexapod robots have been constructed according to varying technologies and design principles. Several hexapod robots limited to 14 kg at maximum are listed in Table 4.

Table 4. A variety of hexapod robots.

Year	Robot Name	Actuator No.	Power (W)	Mass (Kg)	Length (mm)	Width (mm)	Height (mm)	Remarks
2000	Biobot	18	NA [1]	11	580	140	230	Locomotion over rough terrain
2001	Hamlet	18	52	13	400	280	400	Force and position control
2001	Rhex	6	100	7	530	200	150	Reduced actuators
2002	Sprawlita	12	NA	0.27	160	NA	NA	Inspired on cockroaches
2004	Genghis	12	NA	1	400	150	NA	Developing of a reactive controller
2005	Bill-Ant-p	18	25	2.3	470	330	160	Bio-inspired robot
2006	Gregor I	16	25	1.2	300	90	40	Inspired on cockroaches
2006	RiSE	12	NA	3.2	410	NA	NA	Hexapod climbing robot
2017	Hexapod-I	3	7	4.5	712	641	189	Minimal number of actuators
2018	Hexapod-II	3	35	14	775	694	512	Minimal number of actuators

[1] NA: not available.

Biobot [29], inspired to the cockroach, was a hexapod with a great speed and agility. Hamlet [30] was built to study force and position control on uneven terrain. RHex [31], designed in mechanical simplicity and having only six actuators, can achieve fast and robust forward locomotion. Sprawlita [32] was based on functional principles derived from biomechanical studies of the cockroach to achieve a fast run. Genghis [33] presented an incremental method for building robot control systems analogous to evolution of the organism, i.e., each one being a strict augmentation of the previous one, which control a six legged walking machine. Bill-Ant-p [34], based on ants' behavior, was developed and consisted of 18 degrees of freedom for locomotion, i.e., each leg with three actuators and one force-sensing foot. Gregor I [35] is also a bio-inspired hexapod where the locomotion control is based on the CPG and the core of its control architecture is the Cellular-Nonlinear-Network CPG chip. RiSE [36] is a hexapod robot which is able to climb on a variety of vertical surfaces. It employs arrays of miniature spines to catch on surface asperities.

The hexapod robot of this article, named Hexapod-II, is an improved version of the Hexapod-I [37]. The difference between Hexapod-I and Hexapod-II lies in their biomimetic legs. The former uses four-bar linkages, while the latter uses the TJLs. The four-bar linkage can only generate a circular orbit, which leads the hexapod robot to move in a way with a larger fluctuation (Figure 18) because its step height is always one half of its step size, as two tripod gaits are alternated with a phase lag of 180°. In contrast, the TJL generates an ellipse-like trajectory which possesses a transversal axis longer than the lateral axis, so that it can achieve energy efficient walking, as compared to the circular orbit generated by the four-bar linkage.

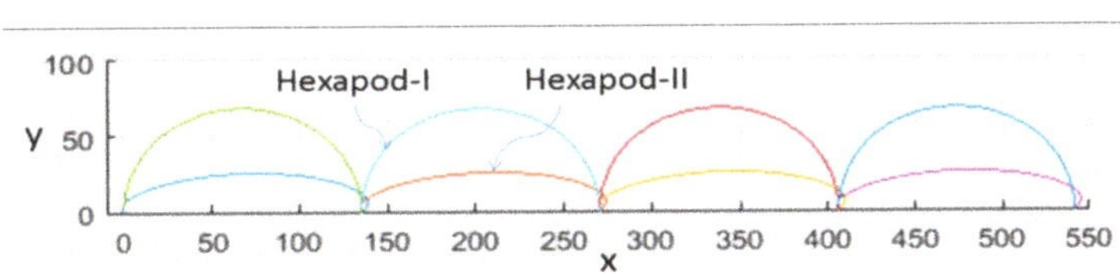

Figure 18. Loci of center of gravity: Hexapod-I versus Hexapod-II.

Note that the cost of a robot is generally proportional to the number of actuators. Therefore, the cost can be roughly estimated by counting the number of actuators. Taking a glimpse at Table 4, the numbers of actuators used by other hexapods are all much higher than the number of actuators used by our robots, i.e., Hexapod-I and Hexapod-II. This is due to the novelty of our approach, which uses the mechanical mechanism rather than electronic circuits or computers to act as the CPG in achieving the locomotion.

7. Conclusions

According to neuroscience, a distributed control system is proposed for the locomotion of a legged robot, since the CPG, located in the spinal cord and in charge of the rhythmic motion, is an autonomous device, almost requiring neither the peripheral sensor feedback nor the regulation command from the brain-stem. A hexapod robot with biomimetic legs was built to realize such a distributed control system, where a mechanism is proposed to serve as the CPG and a computer to act as the brain-stem, wirelessly sending commands to the autonomous device. The proposed mechanism consists of two modules, i.e., the tripod gait generator and the Theo Jansen Linkage. The tripod gait generator is a device that uses a single motor to generate a tripod gait, while the TJL rhythmically executes the legged motion. The interesting point is that the complex mathematical function of the foot motion can be realized by the ensemble of links of the TJL. If the same function is executed by an electronic computer, it will require a great amount of the computational time and memory resources. Nevertheless, dimensioning the TJL appropriately is nontrivial, since the patterns of orbits generated by a TJL may vary according to its overall dimensions and can be cast into four groups: bell curves, ovals, sharp-pointed ovals, and lemniscates. In general, the ovals or bell orbits are legitimate, the sharp-pointed ovals are partly legitimate, while the lemniscates are illegitimate. Once the dimension of a TJL has been determined, it can generate an ellipse-like orbit with a transversal axis longer than the lateral axis, so as to achieve energy efficient walking.

Admittedly, the mechanical computer is inferior to the electronic computer in regards to flexibility. Therefore, in the future, efforts should be taken introduce the adjustable mechanical structure to increase the adaptability of this line of approach. Although the proposed method is not suitable for the case of a legged robot, adapted to diversified terrain, it suits the case of a legged robot operating in a specific environment with relative unevenness, such as construction sites, where the grounds are somewhat muddy and rugged, or factories, where the floors are scattered with debris. Our aim is to build a hexapod robot for material handling, since using a legged robot as a mobile platform is still more effective than using a wheeled robot with regards to crossing obstacles and avoiding skidding.

Robots are, in fact, complex and expensive machines, consisting of many actuators, sensors, transmissions, and hardware. Designing a robot capable of especially excessive function can further increase its cost. The cost of building a robot is generally proportional to the number of actuators it uses. For instance, most hexapod robots, designed with collocated actuators, require 18 servos. Based on the design of non-collocated actuators, the proposed hexapod robot uses merely three motors. The value of this approach lies on the fact that it gives a way of building a hexapod robot with much low cost. The significance of this research is in employing a mechanical mechanism, rather than electronic circuits or computers, to act as the CPG in achieving locomotion.

Author Contributions: M.-C.H. designed the mechanism and F.L. built the mechatronic control system. J.Y. analyzed the kinematics of the Theo Jansen linkage and Y.L. developed the programs. M.-C.H. and F.L. contributed to realization and revision of the manuscript.

Funding: This research was funded by the Jiangxi University of Science and Technology, People's Republic of China (Grant No. jxxjbs18018).

Conflicts of Interest: The authors declare no conflicts of interest.

References

1. McGhee, R.B. Some finite state aspect of legged locomotion. *Math. Biosci.* **1968**, *2*, 67–84. [CrossRef]
2. McGhee, R.B.; Frank, A.A. On the stability properties of quadruped creeping gaits. *Math. Biosci.* **1968**, *3*, 331–351.
3. Orin, D. Interactive Control of a Six-Legged Vehicle with Optimization of Both Stability and Energy. Ph.D. Thesis, Ohio State University, Columbus, OH, USA, 1976.

4. Vukobratovic, M.; Juricic, D. Contribution to the synthesis of biped gait. *IEEE Trans. Biomed. Eng.* **1969**, *BME-16*, 1–6. [CrossRef]
5. Vukobratovic, M.; Borovac, B. Zero-moment point—Thirty five years of its life. *Int. J. Hum. Robot.* **2004**, *1*, 157–173. [CrossRef]
6. Bässler, U. On the Definition of Central Pattern Generator and its Sensory Control. *Biol. Cybern.* **1986**, *54*, 65–69. [CrossRef]
7. Matsuoka, K. Mechanism of Frequency and Pattern Control in the Neural Rhythm Generators. *Biol. Cybern.* **1987**, *56*, 345–353. [CrossRef] [PubMed]
8. Collins, J.J.; Richmond, S.A. Hard–Wired Central Pattern Generators for Quadrupedal Locomotion. *Biol. Cybern.* **1994**, *71*, 375–385. [CrossRef]
9. Ijspeert, A.J. Central Pattern Generators for Locomotion Control in Animals and Robots: A Review. *Neural Netw.* **2008**, *21*, 642–653. [CrossRef]
10. Yu, H.; Gao, H.; Ding, L.; Li, M.; Deng, Z.; Liu, G. Gait Generation with Smooth Transition Using CPG-Based Locomotion Control for Hexapod Waking Robot. *IEEE Trans. Ind. Electron.* **2016**, *63*, 5488–5500. [CrossRef]
11. Zhong, G.; Chen, L.; Jiao, Z.; Li, J.; Deng, H. Locomotion Control and Gait Planning of a Novel Hexapod Robot Using Biomimetic Neurons. *IEEE Trans. Control Syst. Technol.* **2018**, *26*, 624–636. [CrossRef]
12. Olsen, S.R.; Wilson, R.I. Cracking neural circuits in a tiny brain: New approaches for understanding the neural circuitry of Drosophila. *Trans. Trends Neurosci.* **2008**, *31*, 512–520. [CrossRef] [PubMed]
13. Ijspeert, A.J.; Crespi, A.; Cabelguen, J.M. Simulation and Robotics Studies of Salamander Locomotion. *Neuroinformatics* **2005**, *3*, 171–195. [CrossRef] [PubMed]
14. Aminzare, Z.; Srivastava, V.; Holmes, P. Gait transitions in a phase oscillator model of an insect central pattern generator. *SIAM J. Appl. Dyn. Syst.* **2018**, *17*, 626–671. [CrossRef]
15. Schilling, M.; Hoinville, T.; Schmitz, J. Walknet, a bio-inspired controller for hexapod walking. *Biol. Cybern.* **2013**, *107*, 397–419. [CrossRef]
16. Owaki, D.; Goda, M.; Miyazawa, S.; Ishiguro, A. A Minimal Model Describing Hexapodal Interlimb Coordination: The Tegotae-Based Approach. *Front. Neurorobotics* **2017**, *11*, 1–13. [CrossRef] [PubMed]
17. Romano, D.; Donati, E.; Benelli, G.; Stefanini, C. A review on animal–robot interaction: From bio-hybrid organisms to mixed societies. *Biol. Cybern.* **2019**, *113*, 201–225. [CrossRef] [PubMed]
18. Romano, D.; Benelli, G.; Stefanini, C. Encoding Lateralization of Jump Kinematics and Eye Use in a Locust via Bio-Robotic Artifacts. *J. Exp. Biol.* **2019**, *222*, jeb187427. [CrossRef]
19. Strandbeest Evolution 2018. Available online: http://www.strandbeest.com/ (accessed on 18 October 2018).
20. Komoda, K.; Wagatsums, H. A Proposal of the extended mechanism for Theo Jansen linkage to modify the walking elliptic orbit and a study of cyclic bas function. In Proceedings of the 7th Annual Dynamic Walking Conference, Pensacola, FL, USA, 21 February 2012; pp. 2–4.
21. Nansai, S.; Rajesh, M.; Iwase, M. Dynamic Analysis and Modeling of Jansen Mechanism. *Procedia Eng.* **2013**, *64*, 1562–1571. [CrossRef]
22. Komoda, K.; Wagatsuma, H. Singular Configurations Analyses of the Modifiable Theo Jansen-like Mechanism by Focusing on the Jacobian determinant-A Finding Limitations to Exceed Normal Joint Range of Motion. In Proceedings of the 2014 IEEE/ASME International Conference on Advanced Intelligent Mechatronics, Besançon, France, 8–11 July 2014; pp. 76–81.
23. Mohsenizadeh, M.; Zhou, J. Kinematic Analysis and Simulation of Theo Jansen Mechanism. In Proceedings of the Fifteenth Annual Early Career Technical Conference, Tuscaloosa, AL, USA, 7–8 November 2015; pp. 97–104.
24. Nansai, S.; Mohan, R.E.; Tan, N.; Rojas, N.; Iwase, M. Dynamic Modeling and Nonlinear Position Control of a Quadruped Robot with Theo Jansen Linkage Mechanisms and a Single Actuator. *J. Robot.* **2015**, *2015*, 1–15. [CrossRef]
25. Nansai, S.; Rajesh Elara, M.; Iwase, M. Speed Control of Jansen Linkage Mechanism for Exquisite Tasks. *J. Adv. Simul. Sci. Eng.* **2016**, *3*, 47–57. [CrossRef]
26. Denavit, J.; Hartenberg, R.S. A Kinematic Notation for Lower-Pair Mechanisms Based on Matrices. *ASME J. Appl. Mech.* **1955**, *77*, 215–221.
27. Arduino. Available online: https://www.arduino.cc/ (accessed on 1 June 2019).
28. Peternella, M.; Salinari, S. Simulation by digital computer of walking machine control system. In Proceedings of the 5th IFAC Symposium on Automatic Control in Space of the Conference, Genova, Italy, 4–8 June 1973.

29. Delcomyn, F.; Nelson, M.E. Architectures for a biomimetic hexapod robot. *Robot. Auton. Syst.* **2000**, *30*, 5–15. [CrossRef]
30. Fielding, M.R.; Dunlop, R.; Damaren, C.J. Hamlet: Force/Position controlled hexapod walker—Design and systems. In Proceedings of the IEEE Conference on Control Applications, Mexico City, Mexico, 5–9 September 2001; pp. 984–989.
31. Saranli, U.; Buehler, M.; Koditschek, D.E. RHex—A simple and highly mobile hexapod robot. *Int. J. Robot. Res.* **2001**, *20*, 616–631. [CrossRef]
32. Cham, J.G.; Bailey, S.A.; Clark, J.E.; Full, R.J.; CutKosky, M.R. Fast and robust: Hexapodal robots via shape deposition manufacturing. *Int. J. Robot. Res.* **2002**, *21*, 869–882. [CrossRef]
33. Brooks, R.A. A robot that walks; emergent behaviors from a carefully evolved network. *Neural Comput.* **1989**, *1*, 253–262. [CrossRef]
34. Lewinger, W.A.; Branicky, M.S.; Quinn, R.D. Insect-inspired, actively compliant hexapod capable of object manipulation. In Proceedings of the 8th International Conference on Climbing and Walking Robots, London, UK, 13–15 September 2005; pp. 65–72.
35. Arena, P.; Fortuna, L.; Frasca, M.; Patane, L.; Pavone, M. Implementation and experimental validation of an autonomous hexapod robot. In Proceedings of the IEEE International Symposium on Circuits and Systems, Kos, Greece, 21–24 May 2006.
36. Asbeck, A.T.; Kim, S.; McClung, A.; Parness, A.; Cutkosky, M.R. Climbing walls with microspines. In Proceedings of the IEEE International Conference Robotics and Automation, Orlando, FL, USA, 15–19 May 2006.
37. Hwang, M.C.; Huang, C.J.; Liu, F.F. A Hexapod Robot with Non-Collocated Actuators. *Appl. Syst. Innov.* **2018**, *1*, 20. [CrossRef]

applied sciences

MDPI

Article

Heating, Ventilation, and Air Conditioning System Optimization Control Strategy Involving Fan Coil Unit Temperature Control

Chang-Ming Lin [1,*], Hsin-Yu Liu [2], Ko-Ying Tseng [2] and Sheng-Fuu Lin [1]

1 Institute of Electrical and Control Engineering, National Chiao Tung University, 1001 University Road, Hsinchu 30010, Taiwan; sflin@mail.nctu.edu.tw

2 Energy and Environment Research Laboratories, Industrial Technology Research Institute, Rm.820, Bldg.51, 8F, 195, Sec.4, Chung Hsing Rd., Chutung, Hsinchu 31040, Taiwan; hyliu@itri.org.tw (H.-Y.L.); optbilly@gmail.com (K.-Y.T.)

* Correspondence: linbarry@itri.org.tw; Tel.: +886-3-591-6187

Received: 2 May 2019; Accepted: 7 June 2019; Published: 11 June 2019

Abstract: The objective of this study was to develop a heating, ventilation, and air conditioning (HVAC) system optimization control strategy involving fan coil unit (FCU) temperature control for energy conservation in chilled water systems to enhance the operating efficiency of HVAC systems. The proposed control strategy involves three techniques, which are described as follows. The first technique is an algorithm for dynamic FCU temperature setting, which enables the FCU temperature to be set in accordance with changes in the outdoor temperature to satisfy the indoor thermal comfort for occupants. The second technique is an approach for determining the indoor cold air demand, which collects the set FCU temperature and converts it to the refrigeration ton required for the chilled water system; this serves as the control target for ensuring optimal HVAC operation. The third technique is a genetic algorithm for calculating the minimum energy consumption for an HVAC system. The genetic algorithm determines the pump operating frequency associated with minimum energy consumption per refrigeration ton to control energy conservation. To demonstrate the effectiveness of the proposed HVAC system optimization control strategy combining FCU temperature control, this study conducted a field experiment. The results revealed that the proposed strategy enabled an HVAC system to achieve 39.71% energy conservation compared with an HVAC system operating at full load.

Keywords: HVAC energy conservation; optimal control; genetic algorithm; FCU group control

1. Introduction

Air conditioning, lighting, and ventilation systems constitute approximately 50% of the total power consumption of a building. In particular, air conditioning systems have a considerable effect on energy consumption; hence, energy conservation control mechanisms for air conditioning systems are crucial [1]. Air conditioning systems in a mass building are typically centralized and comprise fan coil units (FCUs) in each area of the building and a chilled water system in the mechanical room [2]. The chilled water system provides chilled water to the cooling coils of the FCUs, and the water in the coils is subsequently converted to cold air that is distributed to each area of the building. In general, because of the change of seasons, chilled water systems are designed on the basis of the maximum load in summer. Therefore, such systems often operate under a partial load ratio, leading to energy waste due to poor operating efficiency. The coefficient of performance of a chilled water system is a critical parameter in evaluating the performance of chilled water systems. A higher coefficient signifies higher chilled water system performance. Chow et al. [3] proposed using a neural network and genetic algorithm optimization to enhance the operating efficiency of a chilled water system to achieve energy conservation. Browne and Bansal [4] used a regression model to predict the performance of a chilled water system to optimize the control of the system. Numerous studies have focused on efficiency optimization control strategies for chilled water systems and cooling towers. For example, Yu and Chan [5] used load-based speed control to determine the optimal control frequency for cooling towers and chilled water pumps to optimize the operating efficiency of chilled water systems. In addition to achieving optimal efficiency for chilled water systems (by controlling their operation), effectively controlling chilled water systems (by using indoor heat demand responses) is another means of achieving cost reduction. Yoon et al. [6] proposed the use of a dynamic demand response controller to change the temperature settings of a chilled water system and control its loading during periods of peak power consumption.

Wang et al. [7] revealed that thermal comfort is a crucial factor influencing the control of heating, ventilation, and air conditioning (HVAC) systems because a user's thermal comfort demand directly affects HVAC system control. Therefore, effective control of indoor thermal comfort temperature contributes to the maintenance of environmental comfort and the reduction of FCU energy waste. Specifically, an optimization control strategy to balance between HVAC energy consumption and indoor thermal comfort is valuable. The most commonly cited experiments on the human perception of thermal comfort have been performed by Fanger (1982). His analysis indicated the sensation of human thermal comfort by six factors: Four physical variables, including air temperature, mean radiant temperature, air velocity, and air humidity, and two personal variables, including metabolic rate and clothing insulation. Fanger and Toftum [8] proposed the PMV model, which agrees well with high-quality field studies in buildings with HVAC systems, situated in cold, temperature, and warm climates, studied during both summer and winter. Model predictive control (MPC) is comprehensively used to determine indoor HVAC temperature settings and comfortable temperature values by employing optimal control methods to reduce energy consumption and maintain indoor thermal comfort. Castilla et al. [9] proposed a hierarchical control method that entails the use of a non-linear MPC strategy to predict control. The purpose of this method is to maintain indoor comfort while preventing an HVAC system from entering high-load operation and to optimize efficiency during low-load operation. Jazizadeh et al. [10] applied a fuzzy control system for predicting indoor thermal comfort to effectively reduce the daily mean flow of an HVAC system. Chen et al. [11,12] has proposed two MPC systems for determining the optimal solution to balance indoor thermal comfort and chilled water system control; moreover, a model integrating MPC and dynamic thermal sensation could maintain indoor thermal comfort while consuming energy at a rate lower than that of the conventional MPC–PMV model. Human perception of thermal comfort generally varies with climate. Oldewurtel et al. [13] applied a stochastic MPC system for predicting the change in climate to calculate the minimum energy required to achieve indoor thermal comfort. Fong [14] indicated that indoor thermal comfort is not only a perceived response of the human body to environmental comfort but also a demand response (DR) to the cooling ability of HVAC systems. Lin et al. [15] applied an algorithm to solve optimal chiller loading (OCL) problems. ASHRAE [16] has developed an industry consensus standard to describe comfort requirements in buildings. The standard is known as ASHRAE Standard 55-2004 Thermal Environmental Conditions for Human Occupancy. The purpose of this standard is to specify the combinations of indoor thermal environmental factors and personal factors that will produce thermal environmental conditions acceptable to a majority of the occupants within the space. Brager and Dear [17] have proposed the distinction between thermal comfort response in air-conditioned and naturally ventilated buildings and suggested that behavioral adaptation incorporated in conventional heat balance models could only partially explain these differences and that comfort was significantly influenced by people's expectations of the thermal environment. Brager and Dear [18] have presented the relationship between optimum temperatures and prevailing indoor/outdoor temperatures in the study demonstrated that adaptation is at work in buildings with centralized HVAC. The results were interpreted to indicate that occupants of HVAC buildings had become finely tuned to the very narrow range of indoor temperatures being presented by current HVAC practice. However, there is potentially a very high energy cost to maintaining those narrowly defined comfortable thermal conditions. Humphreys and Nicol [19] presented an adaptive model for thermal comfort, and used a fuzzy logic system to fine tune the room temperature. Dear et al. [20] explained all fields of applied research; it is important that the thermal environment of simulations are regularly 'ground-truthed' with real comfort assessments from human subjects in either chamber studies or 'real' building occupants in field studies. Thermal comfort evaluations by human subjects are a superior contribution to knowledge, having longer-lasting value to the research community than simulated comfort evaluations coming out of a comfort model. Alfano et al. [21] pointed out that HVAC engineers are, in practice, interested in the air temperature value required to ensure thermal comfort rather than in a comfort index.

Many studies explore the use of PID controllers to control the energy-saving control of HVAC systems with optimized algorithms or to use climatic factors to predict the RT requirements of HVAC systems [22]. A few papers combine HVAC system control with indoor comfort requirements, Kampelis et al. [23] proposed the Daily Discomfort Score (DDS) to assess demand response HVAC control and thermal comfort in a university building. Most of the abovementioned documents are optimized for ice water systems, or use the simulation method to estimate the comfort of the indoor ice-water system energy consumption comparison and prediction, but do not combine the indoor real heat demand and HVAC power consumption. This paper aims to solve the optimal control strategy of the chilled water system of HVAC buildings and approach the request of the occupants for indoor thermal comfort. Occupants' adjustments of temperature settings are used as feedback to determine the actual refrigeration ton (RT) demand for a chilled water system and a fuzzy logic system continually to determine the cooling air demand for a chilled water system. Finally, the operating point associated with minimum energy consumption per unit RT is derived using a genetic algorithm. The effectiveness of this energy conservation algorithm was tested through a field experiment.

2. Research Background

Most energy conservation methods implemented in cooling systems for a single chilled water system apply frequency down-conversion control to the chilled water system according to the change in outdoor climate; alternatively, such methods adjust the parameter settings of the chilled water system on the basis of the minimum RT demand. However, few methods apply energy conservation control according to indoor heat demands. To meet the indoor thermal comfort demand, users adjust the operating temperature (*Top*) of an HVAC system through an FCU temperature control panel. This demand defines the required RT that the chilled water system must produce.

The present study focuses on an HVAC system optimization control strategy involving FCU temperature control. In this strategy, indoor air information is included in the analysis, and a user's heat demand volume for a chilled water system is determined. The determined heat demand volume is then converted to RT demand for the chilled water system. This strategy can be used to change the chilled water flow and cooling water flow to meet indoor heat demands while ensuring minimum power consumption. The main techniques involved in this strategy are as follows: algorithm for dynamic FCU temperature setting, method for indoor heat demand conversion, and optimization control strategy for energy consumption of single cooling machine of the chilled water system.

2.1. Dynamic FCU Temperature Settings

An FCU distributes cold air produced by a chilled water system to an indoor area. In such an area, the demand for cold air can be determined by observing a user's adjustments of FCU temperature settings. The Daily Discomfort Score (DDS) is developed to assess demand response (DR) and thermal comfort in a building. Baseline and preconditioning scenarios are used to demonstrate the effectiveness of the Daily Discomfort Score in evaluation thermal comfort and demand response HVAC system set-point control [23]. The ASHRAE Standard 55-1992 states that the comfort zone for summer conditions, air temperature to be between 23 °C and 26 °C and relative humidity between 20% and 60%. Surveys of human thermal response in South East Asia produce the Auliciems' equation for estimating thermal neutrality on the basis of the mean monthly dry bulb temperature [24].

In the proposed method, a regression model that includes outdoor temperature and indoor sensing temperature as variables is used to calculate the indoor HVAC temperature:

$$Ta = \alpha \times (Ti) + \beta \times (To) + c \tag{1}$$

where *Ta* is the indoor HVAC temperature, *Ti* is the indoor sensing temperature, *To* is the outdoor temperature, α and β are regression coefficients, and c is a constant.

In Equation (1), the range of the indoor HVAC temperature is not constrained; hence, the temperature may be excessively high when the indoor sensing temperature or outdoor temperature is markedly high. To maintain indoor temperature quality, the indoor HVAC temperature can be corrected using tan–1 and limited to the indoor comfortable range of 23–26 °C. The indoor comfort temperature (*Tc*) can be expressed as follows:

$$Tc = [\tan^{-1}(Ta - 24.5)] + 24.5 \tag{2}$$

An experiment conducted in this study revealed that when *Top* was adjusted, the indoor environmental temperature normally required 10–20 min to approach *Top*. To maintain indoor temperature quality and meet the cooling air request, the proposed algorithm based on dynamic FCU temperature settings automatically sends an HVAC setting temperature (*Ts*) that is within the indoor comfort range every 30 min and calculates the temperature difference between *Tc* and *Top* to determine the indoor thermal comfort (*ITC*), as presented in Equation (3):

$$ITC = Tc - Top \tag{3}$$

The derived *ITC* is then used to correct *Ts* to achieve indoor comfort. The conditions for determining *ITC*′ are as follows:

$$ITC' = \begin{cases} +1 \sim -1,\ Ts = Ts + [\Delta T \times (C^{\gamma})] \\ Others,\ Ts = Top \end{cases} \tag{4}$$

where $\Delta T = Tc - Ti$ and C^{γ} is the power law conversion coefficient.

The cooling efficiency of an HVAC system is subject to numerous factors. In general, *Top* is not equal to *Ti*; in particular, *Ti* is typically higher than *Top* in summer. To maintain indoor comfort, the proposed algorithm determines that users are satisfied with the current environmental temperature when the derived *ITC* value is within ±1 °C. Additionally, a gamma transformation function is adjusted through power law conversion to adjust *Ts*; thus, *Ti* approaches *Tc*. The purpose of power law conversion is to ensure a high output *Ts* when the input ΔT is high, which enables rapid indoor temperature response. When the input ΔT is low, a low *Ts* output is obtained to prevent excessive energy waste.

An *ITC* value that is not within ±1 °C signifies that users are not satisfied with the current environmental temperature. Accordingly, *Top* is adjusted using the FCU temperature control panel. In this situation, the FCU dynamic temperature setting algorithm uses *Top* as *Ts* for the HVAC system to meet the user-demanded temperature. Subsequently, the algorithm recalculates *Tc* and re-executes the *ITC* determination conditions and temperature control in 30 min.

2.2. Indoor Heat Demand Conversion

A conventional chilled water system estimates the required RT by relying solely on the difference in temperature between chilled water entering and that exiting the system. However, such a system cannot determine whether the estimated RT can meet the heat demand of an area. Therefore, a measurement index is required to enable the chilled water system to adjust RT supplies. A conventional FCU provides only *Ts* and *Ti* information. By contrast, this study proposes a technique that integrates the difference between *Ts* and *Ti* with the FCU wind speed to obtain a comfort weight (W) for chilled water system control. Figure 1 illustrates the system framework.

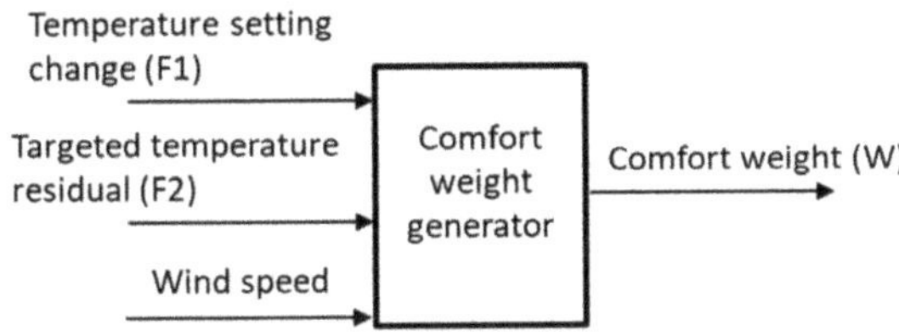

Figure 1. Framework of system for deriving comfort weight.

This system is based on a multiple-input single-output (MISO) fuzzy controller, where the output is W. The MISO controller is described as follows:

1. The change in temperature settings (F1) is used to understand the required RT for a user, as presented in Equation (5). A negative F1 value (e.g., a change in temperature setting from 26 °C to 24 °C) indicates that the user feels hot, whereas a positive F1 value indicates that the user feels cold:

$$F_1 = Ts\ (t) - Ts\ (t-1) \tag{5}$$

   In Equation (5), t is the sampling time.

2. The target temperature error (F2) is the difference between Ti and Ts, and it represents indoor cooling level, as presented in Equation (6). A negative F2 value (e.g., an actual indoor temperature of 24 °C and set temperature of 26 °C) signifies overcooling, whereas a positive F2 value signifies undercooling. Additionally, wind speed is used to determine whether the FCU is turned on or off:

$$\mathrm{F}_2 = \mathrm{Ti} - \mathrm{Ts} \tag{6}$$

   In Equation (6), N represents a negative value and P represents a positive value.

Figure 2 shows the input and output fuzzy sets of the fuzzy controller. Table 1 presents the fuzzy rules applied by the system for weight derivation. A *W* value of 2 means insufficient RT from the chilled water system, indicating a warm indoor environment. By contrast, a *W* value of −2 means excessive RT from the chilled water system, indicating a cold indoor environment.

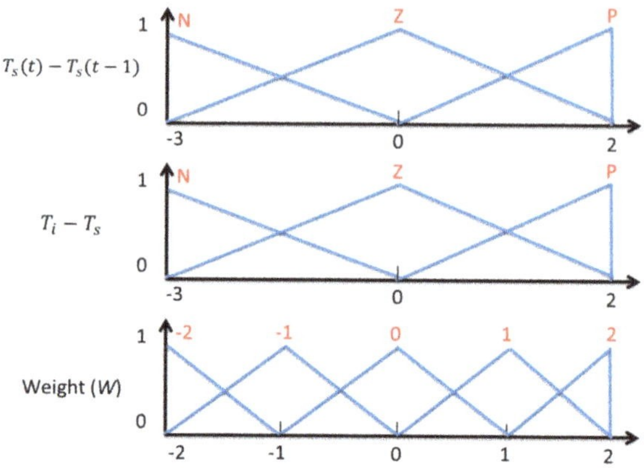

Figure 2. Fuzzy controller.

Table 1. Fuzzy rule.

F2 \ F1	N	Z	P
N	0	−1	−2
Z	1	0	−1
P	2	1	0

In the proposed method, the minimum inference engine and center of gravity defuzzification are used to calculate the comfort weight per time unit *W(t)*, which is presented as follows:

$$W(t) = \frac{\int_T B' t dt}{\int_T B'(t) dt} \tag{7}$$

where *T* denotes the coverage range of sampling time t and B′ denotes the fuzzy set processed by the minimum inference engine.

2.3. Optimization Control Strategy for Single Cooling Machine of Chilled Water System

Among heuristic optimization methods, genetic algorithms are most commonly used. Such algorithms are based on the concept of gene combination in evolution theory. The mechanisms of evolution of life, namely crossover, mutation, and reproduction, are used in genetic algorithms to obtain an optimal solution to a problem. The operating procedures of a genetic algorithm are described below.

2.3.1. Coding

Each individual is randomly assigned a value of 0 or 1 to generate multiple sets of initial solutions. The decoding domains of all initial solutions must fit various design requirements (e.g., the frequency range for a water pump is between 40 and 60 Hz). The solutions are coded using a user operating model, as presented in Figure 3.

Individual 1	1	0	0	0	1	1	0	0	1
Individual 2	0	1	1	0	0	1	1	0	0

Figure 3. Individual gene coding in genetic algorithm.

2.3.2. Fitness Value Calculation and Reproduction Mechanism

Individuals in an initial population are substituted into a target function to calculate a fitness value. The aim of the target function is to obtain the minimum total power consumption of the water system. The total power consumed by the water system is composed of the power consumed by the cooling machine, water pump, and cooling tower. The power consumed by the cooling tower and water pump can be calculated according to the operating frequency, whereas the power consumed by the cooling machine can be calculated using a multiple linear regression model. Such a model can be established by referring to various water system parameters, including chilled water input and output temperatures, chilled water flow, cooling water input and output temperatures, cooling water flow, cooling machine power consumption, water pump frequency, and outdoor temperature and humidity. Furthermore, such a model can be used to estimate the effects of control conditions in different environments on the power consumption of the cooling machines. The reproduction mechanism involves the application of roulette wheel selection, in which each feasible solution is assigned a roulette wheel slot whose area is proportional to the fitness value of the solution. Thus, a solution with a high fitness value has a high probability of being selected. A power consumption model for chilled water systems applies principal component analysis to select the most influential parameters to establish a regression model for evaluating the power consumption of such systems. Such a model is presented in Equation (8):

$$kW_{chi} = a(Tin_{chi}) - b(Tout_{chi}) + c(Tin_{cw}) - d(Tout_{cw}) - e(Tout) - f(Hout) \tag{8}$$

where a, b, c, d, e, and f represent regression parameters; Tin_{chi} represents chilled water input temperature; $Tout_{chi}$ represents chilled water output temperature; Tin_{cw} represents cooling water input temperature; $Tout_{cw}$ represents cooling water output temperature; $Tout$ represent outdoor temperature; and $Hout$ represents outdoor humidity.

2.3.3. Crossover

Crossover entails producing a new generation of "child" solutions by recombining the genes of a reproduced individual and a "parent". A single point is selected as the initial crossover point. Figure 4 shows the genetic algorithm crossover process: the red line indicates the genetic crossover point; for each individual, the genetic data behind the red line are replaced.

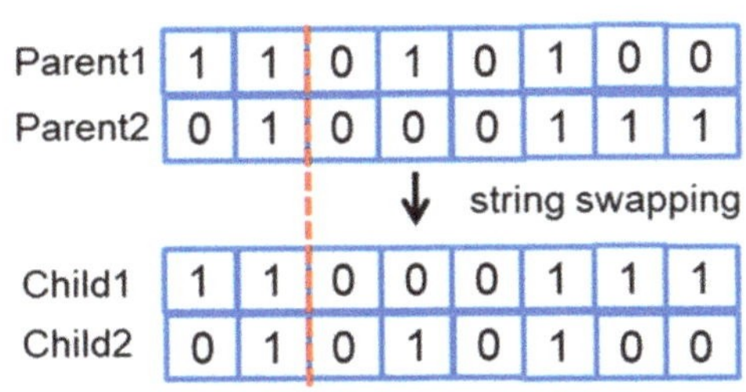

Figure 4. Producing child generations through crossover.

2.3.4. Mutation

Mutation entails randomly selecting a child and replacing one of its genes through a single transformation (replace 0 with 1 or replace 1 with 0). The odds of mutation are considerably lower than those of crossover. Figure 5 illustrates the mutation process.

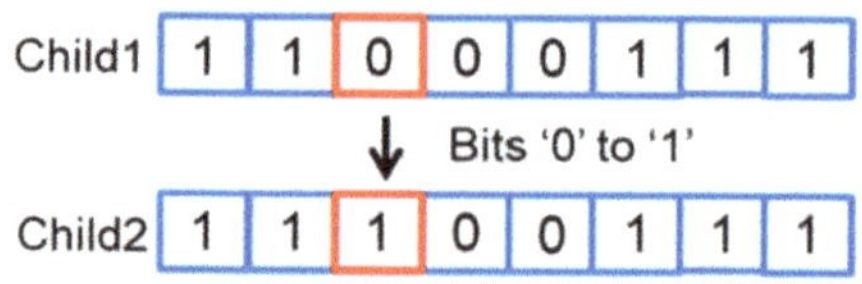

Figure 5. Mutation process of a child.

In summary, the genetic algorithm first executes (A) coding, followed by (B) fitness value calculation and reproduction, (C) crossover, and (D) mutation. A selection process is executed to choose a parent from newly reproduced child individuals, and the fitness value is recalculated to determine the suitability of the selected individual becoming a new parent. This evolution process is repeated until a termination condition is reached and an optimal solution is obtained. Examples of termination conditions are outlined as follows: (1) a limited number of generations is reached, and (2) the optimal solution is not updated for a certain number of consecutive iterations. The search process is considered to have reached convergence when either condition is met. Figure 6 shows the overall operating procedure of the genetic algorithm.

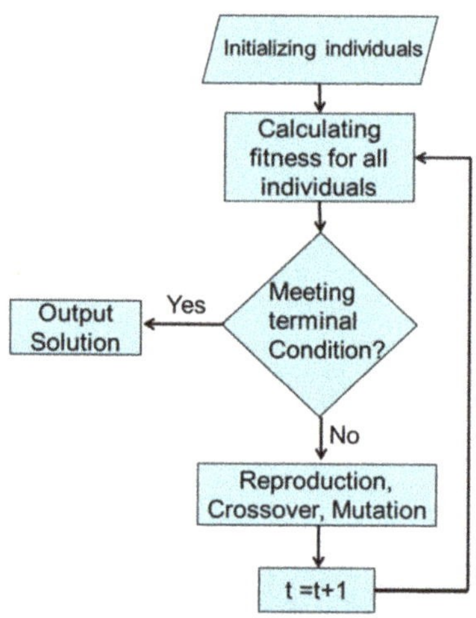

Figure 6. Genetic algorithm flowchart.

The parameters adjusted for the applied genetic algorithm are as follows: chilled water flow, cooling water flow, and chilled water output temperature. The calculated flow is converted to chilled water pump and cooling water pump frequencies by using similar principles. Finally, the chilled water temperature is set. Thus, the three parameters are controlled simultaneously.

3. Case Study

This study executed the proposed HVAC system optimization control strategy involving FCU temperature control in a university in Taipei City, Taiwan. The study focused on energy conservation optimization control for an HVAC system in the university to meet the required levels of indoor thermal comfort. The HVAC system was installed in the teaching and administration building of the university. This HVAC system was determined to comprise a chilled water system, a chilled water pump, a cooling water pump, and two cooling towers. The teaching and administration building was observed to be four stories high, with the HVAC system being operated from 08:00 to 18:00 from Mondays to Fridays; 42 FCUs supply cold air. This study used a Gateway communication system to monitor the water system operation status of the HVAC in the building. In addition, supervisory control and data acquisition (SCADA) was used for energy conservation control in the centralized HVAC system.

The SCADA functions are outlined as follows:

1. FCU data collection

Data, such as *Ti*, *Ts*, and wind speed of the 42 FCUs in the teaching and administration building, were stored in a back-end server using the Modbus transmission method.

2. Chilled water system power consumption monitoring

Digital electricity meters were installed on the chilled water system, chilled water pump, cooling pump, and cooling towers of the HVAC system. Power consumption data were wirelessly transmitted to the SCADA database.

3. Chilled water system flow and temperature monitoring

Flow meters were installed on the piping system of the chilled water system and cooling machine, and thermometers were installed on the outlet and return pipes of the chilled water system. Therefore, data concerning the flow of chilled and cooling water and the temperatures of chilled and cooling water in the outlet and return pipes of the HVAC system were transmitted to the PLC controller. Finally, the entire sensor-driven data of the HVAC system were obtained through SCADA using the Modbus transmission method.

4. Chilled water system ON/OFF and frequency change control

Frequency converters were installed on the chilled water pump, cooling water pump, and cooling tower fans of the chilled water system. Such converters can adjust the frequencies of the chilled water pump, cooling water pump, and cooling tower fans in the event of changes in heat load. Moreover, they can turn the setups on and off through automatic scheduling.

5. FCU temperature control

The FCU temperature control point in the teaching and administration building enabled automatically setting the indoor temperature according to dynamic FCU temperature settings. Such control was achieved by transmitting temperature settings through SCADA.

The HVAC system in the aforementioned building was operated at full load (i.e., 60 Hz) in 2016. The annual mean power consumption of the chilled water system was 33.1 kWh. During the summer (i.e., June, July, and August) of that year—characterized by high power consumption—the monthly mean power consumption of the chilled water system was 62.7 kWh. Figure 7 shows statistics of the annual outdoor temperature distribution and power consumption of the HVAC system.

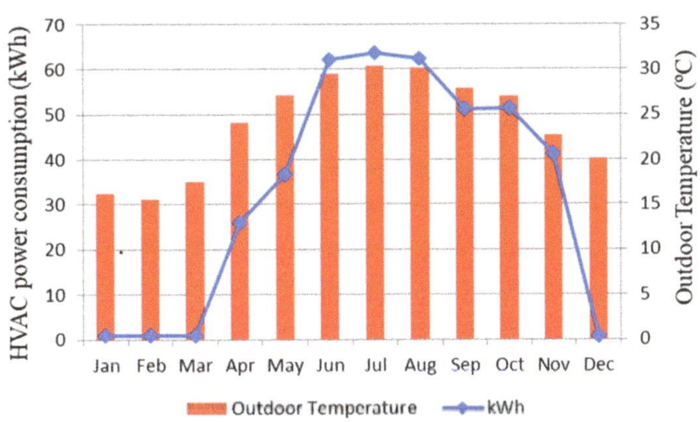

Figure 7. Outdoor temperature and HVAC system power consumption.

In general, outdoor temperature is a crucial factor affecting the power consumption of an HVAC system in a building. Figure 8 illustrates a regression model of the outdoor temperature and power consumption of the chilled water system in 2016. The reliability of this regression model was determined to be 90.9%. As shown in this figure, the power consumption of the chilled water system increased as the outdoor temperature increased. The regression equation for power consumption is presented as follows:

$$kWh = 4.56x - 76.179, \tag{9}$$

where the regression coefficient $x = To$.

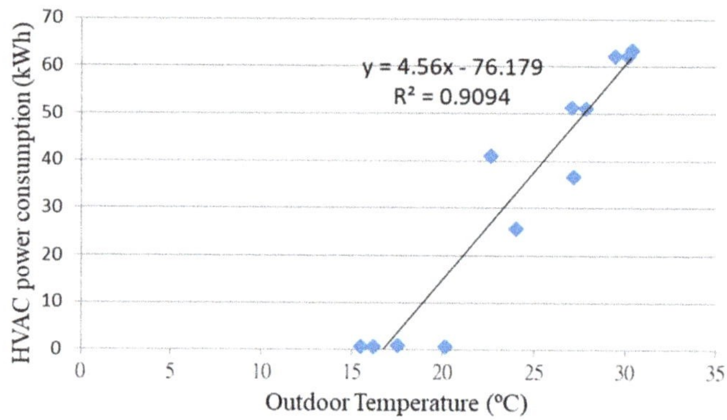

Figure 8. Energy consumption regression model of the chilled water system.

This study also applied the proposed HVAC system optimization control strategy involving FCU temperature control to a single chilled water system with variable frequency from June–August 2017. This strategy was used to calculate the heat demand of the building to correct the supply of chilled water flow. Compared with the HVAC system operated at full load, the HVAC system operated at optimized settings provided by the proposed strategy achieved maximum energy conservation in August, saving 54.9% of energy. The maximum efficiency (in kW/RT) of the chilled water system was 15.6%. Tables 2–4 present comparisons of the performance of the HVAC system operated at full load with that of the system operated at the optimized load in terms of power consumption, RT produced, and chilled water system efficiency in summer (i.e., from June to August).

Table 2. Data of HVAC system operated at full load.

Full-Load Operation	June	July	August
RT	176,550.7	378,983.8	413,532.6
kW	154,125	331,933.6	289,985.9
kW/RT	0.872979	0.875852	0.701241

Table 3. System optimization control strategy involving FCU temperature information.

Optimal Control Operation	June	July	August
RT	253,698.9	247,613.1	191,023.2
kW	170,202.6	166,792.5	130,912.7
kW/RT	0.670885	0.673601	0.685324

Table 4. Energy consumption comparison.

Comparison	RT	kW	kW/RT
Full load	969,067.1	776,044.5	0.800816
Optimization	692,335.1	467,907.9	0.67584
Energy conservation percentage	28.56%	39.71%	15.61%

Figure 9 shows a comparison of outdoor temperatures in 2016 with those in 2017. To understand the energy conservation engendered by the proposed strategy, the outdoor temperatures in 2017 were substituted into Equation (9)—the regression model for calculating the power consumption of a chilled water system—to determine the energy conservation effectiveness associated with various outdoor temperatures. The actual power consumption engendered by the proposed strategy was compared with the baseline power consumption calculated using the regression equation. The comparison revealed a maximum energy conservation of 65.6% (Figure 10); the energy conservation increased when outdoor temperature increased. Therefore, the proposed strategy can achieve energy conservation.

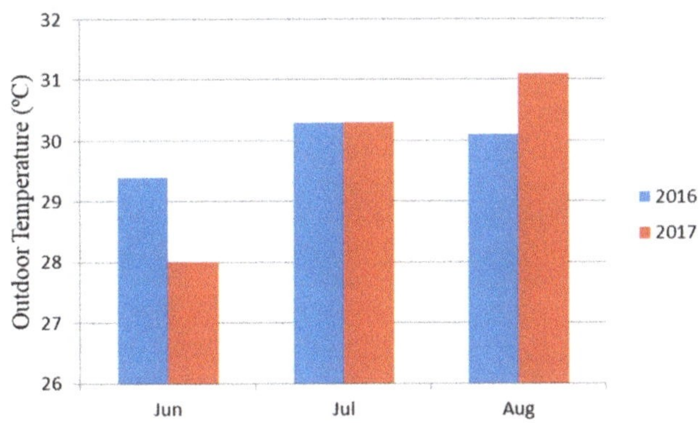

Figure 9. Outdoor temperature distribution.

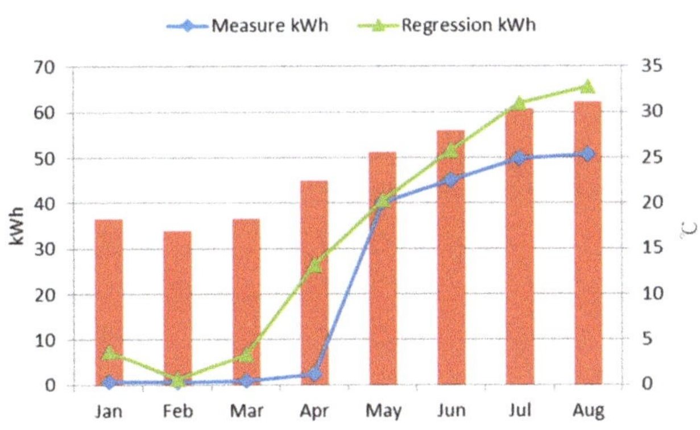

Figure 10. Energy consumption simulation of HVAC system regression model.

4. Conclusions

This study proposes an HVAC system optimization control strategy involving FCU temperature control. This strategy involves three main techniques: dynamic FCU temperature setting, indoor heat demand conversion, energy consumption optimization control strategy for a single chiller unit of HVAC systems. The algorithm for dynamic FCU temperature setting estimates a comfortable indoor temperature according to outdoor and sensed indoor temperatures, and automatically adjust the difference between the indoor comfort temperature (*Tc*) and the indoor sensing temperature (*Ti*) every 30 minutes, and achieve the indoor comfort temperature (*Tc*) within the indoor comfortable range (23–26 °C). The difference in temperature is derived and then converted into an RT for a HVAC system by using fuzzy rules and weight rules on the basis of the change the temperature and wind speed of the FCU by the user. The derived RT is considered the actual cooling air demand of an area for the chilled water system. A genetic algorithm is then applied to adjust chilled water and cooling water flow and calculate the operating frequency of the pumps that are required to meet the cooling air demand and stipulated minimal energy consumption. The effectiveness of the proposed strategy was verified in a field experiment. The study compared the operating status levels of the chilled water pump, cooling water pump, and cooling tower fans in the chosen building during the summer periods (i.e., from June–August) of 2016 and 2017. In 2016, the machines operated at full load, whereas in 2017, the proposed strategy was applied, which enabled the adjustment of the operating frequencies of chilled water system and cooling water pump and airflow of the cooling tower fans of the HVAC system to reduce unnecessary power consumption. Therefore, compared with that in 2016, the actual total power consumption of the HVAC water system in 2017 was reduced by 39.7%, on average. The operating efficiency of the chilled water system was enhanced to 0.68 from 0.8 observed when the system operated at full load. Energy consumption per RT was reduced by approximately 15.6%. According to the baseline energy consumption obtained by a regression model used to calculate the power consumption of a chilled water system, the proposed strategy can achieve energy conservation when temperatures increase.

Author Contributions: Formal analysis, H.-Y.L.; Methodology, C.-M.L.; Software, K.-Y.T.; Validation, S.-F.L.

Funding: Bureau of Energy, Ministry of Economic Affairs, Taiwan: 108-E0203.

Acknowledgments: We appreciate National Chiao Tung University for their generous assistance of computing time and Bureau of Energy, Ministry of Economic Affairs, R.O.C. (Taiwan) for the finical support.

Conflicts of Interest: The authors declare no conflict of interest.

References

1. Pe'rez-Lombard, L.; Ortiz, J.; Pout, C. A review on buildings energy consumption information. *Energy Build.* **2008**, *40*, 394–398. [CrossRef]
2. Mathews, E.H.; Botha, C.P.; Arndt, D.C.; Malan, A. HVAC control strategies to enhance comfort and minimize energy usage. *Energy Build.* **2001**, *33*, 853–863. [CrossRef]
3. Chow, T.T.; Zhang, G.Q.; Lin, Z.; Song, C.L. Global optimization of absorption chiller system by genetic algorithm and neural network. *Energy Build.* **2002**, *34*, 103–109. [CrossRef]
4. Browne, M.W.; Bansal, P.K. Transient simulation of vapour-compression packaged liquid chillers. *Int. J. Refrig.* **2002**, *25*, 597–610. [CrossRef]
5. Yu, F.W.; Chan, K.T. Optimization of water-cooled chiller system with load-based speed control. *Appl. Energy* **2008**, *85*, 931–950. [CrossRef]
6. Yoon, J.H.; Baldick, R.; Novoselac, A. Dynamic demand response controller based on real-time retail price for residential buildings. *IEEE Trans. Smart Grid* **2014**, *5*, 121–129. [CrossRef]
7. Wang, Y.; Kuckelkorn, J.; Zhao, F.-Y.; Liu, D.; Kirschbaum, A.; Zhang, J.-L. Evaluation on classroom thermal comfort and energy performance of passive school by optimizing HVAC control systems. *Build. Environ.* **2015**, *89*, 86–106. [CrossRef]
8. Fanger, P.O.; Toftum, J. Extension of the PMV model to non-air-conditioned buildings in warm climates. *Energy Build.* **2002**, *34*, 533–536. [CrossRef]
9. Castilla, M.; Álvarez, J.D.; Normey-Rico, J.E.; Rodríguez, F. Thermal comfort control using a non-linear MPC strategy: A real case of study in a bioclimatic building. *J. Process Control* **2014**, *24*, 703–713. [CrossRef]
10. Jazizadeh, F.; Ghahramani, A.; Becerik-Gerber, B.; Kichkaylo, T.; Orosz, M. User-led decentralized thermal comfort driven HVAC operations for improved efficiency in office buildings. *Energy Build.* **2014**, *70*, 398–410. [CrossRef]

11. Chen, X.; Wang, Q.; Srebric, J. Model predictive control for indoor thermal comfort and energy optimization using occupant feedback. *Energy Build.* **2015**, *102*, 357–369. [CrossRef]
12. Chen, X.; Wang, Q.; Srebric, J. Occupant feedback based model predictive control for thermal comfort and energy optimization: A chamber experimental evaluation. *Appl. Energy* **2016**, *164*, 341–351. [CrossRef]
13. Oldewurtel, F.; Parisio, A.; Jones, C.N.; Gyalistras, D.; Gwerder, M.; Stauch, V.; Lehmann, B.; Morari, M. Use of model predictive control and weather forecasts for energy efficient building climate control. *Energy Build.* **2012**, *45*, 15–27. [CrossRef]
14. Fong, K.F.; Hanby, V.I.; Chow, T.T. HVAC system optimization for energy management by evolutionary programming. *Energy Build.* **2006**, *38*, 220–231. [CrossRef]
15. Lin, C.-M.; Wu, C.-Y.; Tseng, K.-Y.; Ku, C.-C.; Lin, S.-F. Applying two-stage differential evolution for energy saving in optimal chiller loading. *Energies* **2019**, *12*, 622. [CrossRef]
16. ASHRAE 55. *Thermal Environmental Conditions for Human Occupancy*; American Society of Heating, Refrigerating, and Air-Conditioning Engineers, Inc.: Atlanta, GA, USA, 2017.
17. Brager, G.S.; de Dear, R.J. Thermal adaptation in the built environment: A liter-ature review. *Energy Build.* **1998**, *27*, 83–96. [CrossRef]
18. Brager, G.S.; de Dear, R.J. Developing an adaptive model of thermal comfort and preference. *ASHRAE Trans.* **1998**, *104*, 145–167.
19. Humphreys, M.A.; Nicol, J.F. Understanding the adaptive approach to thermal comfort. *ASHRAE Trans.* **1998**, *104*, 991–1000.
20. De Dear, R.J.; Akimoto, T.; Arens, E.A.; Brager, G.; Candido, C.; Cheong, K.W.D.; Li, B.; Nishihara, N.; Sekhar, S.C.; Tanabe, S.; et al. Progress inthermal comfort research over the last twenty years. *Indoor Air* **2013**, *23*, 442–461. [CrossRef]
21. Alfano, F.R.d.; Olesen, B.W.; Palella, B.I. Povl Ole Fanger's Impact Ten Years Later. *Energy and Buildings* **2017**, *152*, 243–249. [CrossRef]
22. Uipiani, G.; Borgognoni, M.; Romagnoli, A.; di Perna, C. Comparing the performance of on/off, PID and fuzzy controllers applied to the heating system of an energy-efficient building. *Energy Build.* **2016**, *116*, 1–17. [CrossRef]
23. Kampelis, N.; Ferrante, A.; Kolokotsa, D.; Gobakis, K.; Standardi, L.; Cristalli, C. Thermal comfort evaluation in HVAC Demand Response control. *Energy Procedia* **2017**, *134*, 675–682. [CrossRef]
24. Daghigh, R.; Adam, N.M.; Sahari, B.; Yousef, B.A.A. Thermal Comfort Study and Ventilation Evaluation of an Office. *Adv. Mech. Eng. Its Appl.* **2012**, *3*, 278–283.

MDPI
St. Alban-Anlage 66
4052 Basel
Switzerland
Tel. +41 61 683 77 34
Fax +41 61 302 89 18
www.mdpi.com

Applied Sciences Editorial Office
E-mail: applsci@mdpi.com
www.mdpi.com/journal/applsci

www.ingramcontent.com/pod-product-compliance
Lightning Source LLC
Chambersburg PA
CBHW062036220926
51850CB00010B/2573
* 9 7 8 3 0 3 9 2 8 9 2 9 5 *